Networked Artificial Intelligence

The integration of fifth-generation (5G) wireless technologies with distributed artificial intelligence (AI) is transforming network operations. AI is increasingly embedded in all network elements, from cloud and edge to terminal devices, enabling AI to function as a networking system. This convergence facilitates AI-based applications across the global network, with notable successes in various domains such as computer vision, natural language processing, and healthcare. *Networked Artificial Intelligence: AI-Enabled 5G Networking* is a comprehensive framework for the deep integration of computing and communications, optimizing networks and applications as a unified system using AI.

The book covers topics ranging from networked AI fundamentals to AI-enabled 5G networks, including agent modeling, machine learning (ML) algorithms, and network protocol architectures. It discusses how network service providers can leverage AI and ML techniques to customize network baselines, reduce noise, and accurately identify issues. Also, it looks at AI-driven networks that enable self-correction for maximum uptime and prescriptive actions for issue resolution, as well as troubleshooting by capturing and storing data before network events.

The book presents a comprehensive approach to AI-enabled networking that offers unprecedented opportunities for efficiency, reliability, and innovation in telecommunications. It works through the approach's five steps of connection, communication, collaboration, curation, and community. These steps enhance network effects, empowering operators with insights for trusted automation, cost reduction, and optimal user experiences. The book also discusses AI and ML capabilities that enable networks to continuously learn, self-optimize, and predict and rectify service degradations proactively, even with full automation.

Networked Artificial Intelligence

Intelligence

AI-Enabled 5G Networking

Radhika Ranjan Roy

CRC Press
Taylor & Francis Group
Boca Raton London New York

CRC Press is an imprint of the
Taylor & Francis Group, an **informa** business

AN AUERBACH BOOK

First edition published 2025
2385 NW Executive Center Drive, Suite 320, Boca Raton FL 33431

and by CRC Press
4 Park Square, Milton Park, Abingdon, Oxon, OX14 4RN

CRC Press is an imprint of Taylor & Francis Group, LLC

© 2025 Taylor & Francis Group, LLC

ISBN: 978-1-032-80389-0 (hbk)
ISBN: 978-1-032-81369-1 (pbk)
ISBN: 978-1-003-49946-6 (ebk)

DOI: 10.1201/9781003499466

Typeset in Times
by Apex CoVantage, LLC

This book is dedicated to our beloved grandson Amor.

To my Grandma for her causeless love, my parents Rakesh Chandra Roy and Sneholota Roy whose spiritual inspiration remains vividly alive within all of us, my late sisters GitaSree Roy, Anjali Roy, and Aparna Roy and their spouses and my brother Raghunath Roy and his wife Nupur for their inspiration, my daughter Elora and my son-in-law Nick, my son Ajanta, daughter-in-law Nageen and their son and our grandson Amor, and finally my beloved wife Jharna for their love.

To our dearest son, Debasri Roy, Medicinae Doctoris (MD) (January 20, 1988–October 31, 2014) who had a brilliant career (summa cum laude, undergraduate) and had so much more to contribute to this country and the world. He had been so eager to see this book published, saying, "Daddy, you are my hero." Pictured with his fiancé, "Love is forever," he wrote as his lifelong wishes: "I am ready for my life, I like to see the whole world, I will read spiritual scriptures to find the mystery of life, and I want to make a difference in the world." He loved all of us from the deepest part of his kindest heart, including his long-time fiancé who was also an MD and his classmate, and to whom he was engaged and wanted to be married on October 30, 2015, in front of relatives, friends, colleagues, and neighbors who, until his last breath, he had inspired in so many ways to love God, along with his prophetic words: "Mom, I am extremely happy that you are with me and I want to be happy in life." May God let his soul live in peace in His abode.

Contents

Preface

I have been working on networked artificial intelligence for a long time at my present position at United States, Engineering & Systems Integration Directorate, Command Control, Computers, Communications, Cyber, Intelligence, Surveillance and Reconnaissance (C5ISR) Center for artificial intelligence, machine learning, and deep learning (AI/ML/DL) for large-scale global networks, a term coined with networked AI, since 2009. With the advent of fifth generation (5G) wireless technologies. Distributed AI, implemented in network devices, is becoming immersive in all elements of the network, for example, cloud, edge, and terminal devices, which make AI virtually operate as a networking system. 5G technologies are enabling AI-based applications over the global network. AI has achieved great success in areas including computing vision, natural language processing, multimedia production, medical and medicine applications, and even human genetics among others. However, this book has provided a unified framework for the deep convergence of computing and communications, where the network and application/service can be jointly optimized as a single integrated system using AI. This is the first book that deals with full treatment of networked AI.

Network service providers can customize the network baseline for alerts, reducing noise, and false positives while enabling information technology (IT) teams to accurately identify issues, trends, anomalies, and root causes. AI/ML/DL techniques are also used to reduce unknowns and improve the level of certainty in decision making. Networked AI/ML/DL even can enable IT systems to self-correct for maximum uptime and provide prescriptive actions as to how to fix problems that occur. In addition, AI-driven networks can capture and save data prior to a network event or outage, helping to speed up troubleshooting. Connection, communication, collaboration, curation, and community are the five steps that help to boost network effects. Network operators gain insights through analytics and AI/ML/DL that guide more trusted automation processes that lower the cost of network operations and provide users with an optimal connected experience. One important thing is that AI/ML/DL will increasingly enable networks to continually learn, self-optimize, and even predict and rectify service degradations before they occur even with full automation.

This book has 18 chapters:

1. Networked Artificial Intelligence
2. Artificial Intelligent Agent
3. Agent Function
4. Agent Modeling
5. Multi-Agent System
6. Protocol Layer Architecture
7. Artificial Intelligence Performance Analysis
8. Unsupervised Machine Learning
9. Supervised Machine Learning
10. Deep Learning
11. Overfitting and Underfitting
12. Hybrid Learning
13. Reinforcement Learning
14. Artificial Intelligence Application and Network Protocol Architecture Model
15. AI-enabled Network
16. AI-enabled End-to-End Network
17. AI-enabled Peer-to-Peer Network
18. Artificial Intelligence-Enabled 5G Network

The descriptions of all these technologies in detail is a huge undertaking. The book has covered each of the above topics succinctly as far as possible. Each of these sections and subsections are being covered with the texts of from the latest renown published research and development papers. I am deeply indebted to these researchers who have published these papers. I have provided a list of references in each chapter with the names and date of publication.

Author

Radhika Ranjan Roy, PhD, is electronics engineer (Research), Tactical Network Protection Branch, Cyber Security Division Engineering & Systems Integration Directorate, DEVCOM C5ISR Center, Aberdeen Proving Ground (APG), Maryland, since 2009. Dr. Roy has been leading research works on networked artificial intelligence for a long time at my present position at United States, Engineering & Systems Integration Directorate, Command Control, Computers, Communications, Cyber, Intelligence, Surveillance and Reconnaissance (C5ISR) Center for artificial intelligence, machine learning, and deep learning (AI/ML/DL) for large-scale global networks, a term coined with networked AI, since 2009.

Dr. Roy is also leading his research and development efforts in the development of scalable large-scale SIP-based VoIP/multimedia networks and services, mobile ad hoc networks (MANETs), peer-to-peer (P2P) networks, cyber security detection application software and network vulnerability, jamming detection, and supporting an array of US Army/Department of Defense's Nationwide and Worldwide Warfighter Networking Architectures and participating in technical standards development in multimedia/real-time services collaboration, IPv6, radio communications, enterprise services management, and information transfer of Department of Defense (DoD) technical working groups.

He received his PhD in electrical engineering with a major in computer communications from The City University of New York, New York, in 1984 and MS in electrical engineering from Northeastern University, Boston, Massachusetts, in 1978. He received his BS in electrical engineering from Bangladesh University of Engineering and Technology, Dhaka, Bangladesh, in 1967. He was born in the countryside renowned town of Derai, Bangladesh.

Prior to joining C5ISR, Dr. Roy worked as the lead systems engineer at CACI, Eatontown, New Jersey, from 2007 to 2009, and developed Army Technical Resource Model (TRM), Army Enterprise Architecture (AEA), DoD Architecture Framework (DoDAF), and Army LandWarNet (LWN) Capability Sets, as well as technical standards for Joint Tactical Radio System (JTRS), Mobile IPv6, MANET, and Session Initiation Protocol (SIP) supporting Army Chief Information Officer (CIO)/G-6. Dr. Roy worked as a senior systems engineer, SAIC, Abingdon, Maryland, from 2004 to 2007, supporting modeling, simulations, architectures, and system engineering of many Army projects: WIN-T, FCS, and JNN.

During his career, Dr. Roy worked at AT&T/Bell Laboratories, Middletown, New Jersey, as senior consultant from 1990 to 2004, and led a team of engineers in designing AT&T's worldwide SIP-based VoIP/multimedia communications network architecture consisting of wired and wireless parts, from the preparation of Request for Information (RFI), evaluation of vendor RFI responses, and interactions with all selected major vendors related to their products. He participated and contributed to the development of VoIP/H.323/SIP multimedia standards in ITU-T, IETF, ATM, and Frame Relay standard organizations. Dr. Roy worked as senior principal engineer in CSC, Falls Church, Virginia, from 1984 to 1990, and worked in the design and performance analysis of the US Treasury nationwide X.25 packet switching network. In addition, he designed the network architectures of many proposed US government and commercial worldwide and nationwide networks: Department of State Telecommunications Network (DOSTN), US Secret Service Satellite Network, Veteran Communications Network, and Ford Company's Dealership Network. Prior to CSC, he worked from 1967 to 1977 as deputy director (design) in PDP, Dhaka, Bangladesh.

Dr. Roy's research interests include the areas of networked AI, distributed AI, mobile ad hoc networks, multimedia communications, peer-to-peer networking, and quality of service. He has published more than 50 technical papers and is holding or pending more than 30 patents. He is also a member of many IETF working groups. In addition, Dr. Roy has authored three books for

CRC press: *Handbook on Networked Multipoint Multimedia Conferencing and Multi-Stream Immersive Telepresence using SIP*, CRC Press, 2021, *Handbook of SDP for Multimedia Session Negotiations: SIP and WebRTC IP Telephony*, CRC Press, 2018, and *Handbook on Session Initiation Protocol: Networked Multimedia Communications for IP Telephony*, CRC Press, 2016.

He lives in the historical district of Howell Township, New Jersey, with his wife Jharna.

1 Networked Artificial Intelligence

Artificial Intelligence (AI) is playing a major role in the fourth industrial revolution, and we are seeing a lot of evolution in various machine learning methodologies. AI techniques are widely used by practicing engineers to solve a whole range of hitherto intractable problems. Recently, OpenAI has publicly released their primarily large natural language processing (NLP) Model-Based generative AI ChatGPT4. ChatGPT4 can write codes, letters, and poems; plot market strategies; answer questions; pass exams; and do many other things. Moreover, AI is being used in deep learning neural networks and real-world applications; big data analytics and understanding complex networks; Intelligent transportation systems and smart vehicles; Internet-of-things (IoT) and cyber-physical systems; self-organizing and emerging bio-inspired systems; global optimization; evolutionary algorithms; swarm intelligence; nature and biologically inspired meta-heuristics; aspects of reasoning: abductive, case-based, Model-Based, non-monotonic, fuzzy systems; neuro-fuzzy systems; architectures, algorithms, and techniques for distributed AI systems, including multi-agent-based control and holonic control, decision-support systems, incomplete, progressive, and approximate reasoning; applications of chaos theory and fractals; robotics; and many others.

AI not only holds potential benefits and opportunities but also challenges and pitfalls. For example, AI technologies can accelerate and provide insights into data processing, augment human decision-making, optimize performance for complex tasks and systems, and improve safety for people in dangerous occupations. On the other hand, AI systems may perpetuate or amplify bias, may not yet be fully able to explain their decision-making, and often depend on vast datasets that are not widely accessible to facilitate research and development (R&D).

It is observed that there is a general lack of attention to how AI can be used in communications networks to make them AI-enabled, ushering a new era of networked AI. In this book, we are primarily focusing on networked AI that addresses the use of AI in the communications network for making the network AI-enabled. The different components and processes of the communications network such as routers/switches, user equipment (UE), servers, databases, and networking communications protocols will be artificial intelligence, machine learning, and deep learning (AI/ML/DL)-enabled. The AI/ML/DL architecture will specify how data is handled, models are trained and assessed, and predictions are created in satisfying the functional and performance objectives of the communications network.

1.1 EMERGENCE OF AI TECHNOLOGY

Artificial Intelligence (AI) has emerged as the most dominant technology in the 21st century affecting every aspect of all branches of science, technology, and other disciplines as well as human society. AI is hugely complex and consists of a variety of software (SW) and hardware (HW) technologies applied in applications. The emergence of worldwide proprietary AI products and services in the commercial and military arena has created a tremendous push for all companies and governments how interoperability can be provided using standard-based AI-enabled networking architecture and protocols in multi-vendor environments offering scalability and economies-of-scale reducing cost of products and services. The goal of AI is to provide software that can reason on input and explain on output. AI will provide human-like interactions with software and offer decision support for specific tasks, but it's not a replacement for humans [1]. AI augments human intelligence with rich analytics

DOI: 10.1201/9781003499466-1

and pattern prediction capabilities to improve the quality, effectiveness, and creativity of human decisions with improved accuracy.

AI is no longer a standalone product and service. AI services must be provided on a global scale for the entire population of earth and maybe in space as well as connected over the worldwide network like the public internet, commercial, and/or military network. AI products are not only high-end supercomputers for cloud computing, medium-end network-based servers, rouers, and switches of both wide area network and premises networks but also end-user devices like laptops equipped with powerful central processing units (CPUs) and graphical processing units (GPUs) and even handheld tiny smartphones.

To make the matter more complicated, billions of Internet-of-Things (IoT) fueled by explosive growth of fifth-generation (5G) and emerging 6G mobile technologies could also become AI-enabled or would be enabled to have services from AI servers acting as tiny AI-clients. AI-enabled IoT entities are connected over the network and could be very tiny products like sensors, home devices like thermostats, fire alarms, wireless sensor tags, radio-frequency identification (RFID) tags, biometric patches, and others. This new generation of AI-enabled small, tiny IoT entities could not only be fixed but could also be used in mobile environments.

The term artificial intelligence is widely used in the field, and there is much debate how to differentiate other kinds of intelligences (e.g., natural) from AI. We are not discussing this in this book as many confusions remain to be resolved clearly and succinctly, even including Turing and other tests. Rather, we will define in terms of a computer that has some sort of capabilities like, if not equivalent to, human intelligence. So, we define

> Artificial Intelligence (AI) is the field embodied as the intelligent agent that can reason, learn, and act intelligently mimicking human intelligence for problem-solving and decision-making. Detailed planning is needed for before taking any action. We can synthetize and analyze the following components centering on imitating human intelligence: Agent, Reasoning, Learning, Planning, and Acting.

The term artificial intelligence was coined in 1950 by Alan Turing's seminal work, *Computing Machinery and Intelligence*. However, AI has become more popular today thanks to increased data volumes, advanced algorithms, and improvements in computing power and storage. Early AI research in the 1950s explored topics like problem-solving and symbolic methods. In the 1960s, the US Department of Defense took interest in this type of work and began training computers to mimic basic human reasoning. For example, the Defense Advanced Research Projects Agency (DARPA) completed street mapping projects in the 1970s. And DARPA produced intelligent personal assistants in 2003, long before Siri, Alexa, Cortana, or generative AI—ChatGPT were household names. This early work paved the way for the automation and formal reasoning that we see in computers today, including decision support systems and smart search systems that can be designed to complement and augment human abilities. While Hollywood movies and science fiction novels depict AI as human-like robots that take over the world, the current evolution of AI technologies isn't that scary—or quite that smart. Instead, AI has evolved to provide many specific benefits in every industry. Keep reading for modern examples of AI in health care, retail, and more. We are describing how AI has emerged as the most dominant technology of our time because of unprecedented advancement in machine learning (1943–2010), deep learning (2011–2020s)/neural networks (1950s–1970s), and present-day generative AI.

1.1.1 MACHINE LEARNING

Machine learning history starts in 1943 with the first mathematical model of neural networks presented in the scientific paper "A logical calculus of the ideas immanent in nervous activity" by Walter Pitts and Warren McCulloch. However, the field of machine learning was founded by computer scientist Alan Turing in the 1950s. Arthur Samuel is credited with coining the term "machine learning" in 1959 while at IBM. In the years 1950s–2010s, machine learning becomes popular.

Machine learning automates analytical model building. It uses methods from neural networks, statistics, operations research, and physics to find hidden insights in data without explicitly being programmed for where to look or what to conclude.

1.1.2 Deep Learning and Neural Networks

In the years 2011–2020s, breakthroughs in deep learning have driven the AI boom. Deep learning uses huge neural networks with many layers of processing units, taking advantage of advances in computing power and improved training techniques to learn complex patterns in large amounts of data. Common applications include image and speech recognition. Early work in the years 1950s–1970s with neural networks stirs excitement for "thinking machines." A neural network is a type of machine learning that is made up of interconnected units (like neurons) that processes information by responding to external inputs, relaying information between each unit. The process requires multiple passes at the data to find connections and derive meaning from undefined data. In the 1980s, neural networks which use a backpropagation algorithm to train itself became widely used in AI applications. In the year 2015, Baidu's Minwa supercomputer uses a special kind of deep neural network called a convolutional neural network to identify and categorize images with a higher rate of accuracy than the average human.

1.1.3 Generative Artificial Intelligence

Generative AI, a disruptive technology, soars in popularity. For example, ChatGPT being a generative AI that primarily uses large natural language processing (NLP) model has stirred the whole world. It is the big data that makes powerful AI products. So data is said to be the king.

1.2 USAGE OF ARTIFICIAL INTELLIGENCE

- End-to-end network automation with zero-touch provisioning
- Optimized information technology services management like password resets and hardware glitches
- Improved network management and performances identifying bottlenecks, latency issues, and congestion areas
- Faster incident management through correlation analysis, pattern recognition, and other methods
- Intelligent cybersecurity looking at traffic, user behavior, and system logs to pinpoint anomalies and flag potential security breaches or attacks almost in real-time reducing false alarms
- AI-based digital twins that can work into a continuous integration and continuous delivery pipeline allowing "what if" scenarios ensuring that the network is operating as expected
- Accurate image recognition
- Identify products, people, defects, and damages
- Speech recognition
- Natural language generation
- Sentiment analysis
- Chatbots/customer service
- Space exploration
- Health care
- Manufacturing
- Banking
- Retail
- Many others

1.3 ARTIFICIAL INTELLIGENCE ENABLING AND SUPPORTING TECHNOLOGIES

Natural language processing, computer vision, graphical processing units, Internet-of-Things, advanced algorithms, and application programming interfaces are the primary enabling and supporting technologies for AI that have made unprecedented breakthroughs for using AI in all spheres of today's works throughout the world. We will briefly describe these technologies here.

1.3.1 NATURAL LANGUAGE PROCESSING

Natural language processing (NLP) is the ability of computers to analyze, understand, and generate human language, including speech. The next stage of NLP is natural language interaction, which allows humans to communicate with computers using normal, everyday language to perform tasks.

1.3.2 COMPUTER VISION

Computer vision relies on pattern recognition and deep learning to recognize what's in a picture or video. When machines can process, analyze, and understand images, they can capture images or videos in real time and interpret their surroundings.

1.3.3 GRAPHICAL PROCESSING UNITS

Graphical processing units (GPUs) are key to AI because they provide the heavy computer power that's required for iterative processing. Training neural networks requires big data plus compute power.

1.3.4 INTERNET-OF-THINGS

The Internet-of-Things generates massive amounts of data from connected devices, most of it unanalyzed. Automating models with AI will allow us to use more of it.

1.3.5 ADVANCED ALGORITHMS

Advanced algorithms are being developed and combined in new ways to analyze more data faster and at multiple levels. This intelligent processing is key to identifying and predicting rare events, understanding complex systems, and optimizing unique scenarios.

1.3.6 APPLICATION PROGRAMMING INTERFACES

Application programming interfaces (APIs) are portable packages of code that make it possible to add AI functionality to existing products and software packages. They can add image recognition capabilities to home security systems and question and answer capabilities that describe data, create captions and headlines, or call out interesting patterns and insights in data.

1.4 EMERGENCE OF EXTREMELY HIGH-SPEED SEMICONDUCTOR TECHNOLOGIES

AI technology is expected to have a significant impact on the entire industry worldwide. The semiconductor industry is also changing with the development of high-speed semiconductors to meet the requirements of AI-based sophisticated technologies. For example, there are

AI technology, big data technology, 5G wireless communication technology, emerging 6G internet-of-things (IoT) technology, and autonomous driving technology. As a result, the technological capability of semiconductor firms will be upgraded. For example, the development of the neural processing unit (NPU) and graphic processing unit (GPU) with AI technology will create a market for autonomous cars. We are providing a list of emerging high-speed semiconductors as follows:

- Semiconductor Epitaxial Growth (MBE and MOCVD): Metal organic chemical vapor deposition (MOCVD) is a variant of chemical vapor deposition (CVD), generally used for depositing crystalline micro/nano thin films and structures. Fine modulation, abrupt interfaces, and a good level of dopant control can be readily achieved. Molecular beam epitaxy (MBE) is a thin-film deposition process in which thermal beams of atoms or molecules react on the clean surface of a single-crystalline substrate that is held at high temperatures under ultrahigh-vacuum conditions to form an epitaxial film.
- Chemical Vapor Deposition (CVD) Atomic Layer Deposition (ALD): Atomic layer deposition (ALD) is a technique for growing thin films for a wide range of applications. ALD is a special variant of the chemical vapor deposition (CVD) technique where gaseous reactants (precursors) are introduced into the reaction chamber for forming the desired material via chemical surface reactions.
- High-k (k = dielectric constant) dielectrics for ultimate complementary metal-oxide semiconductor (CMOS): High-k dielectric having $k \geq 20$ is beneficial for CMOS nano-devices, reducing the gate leakage current when equivalent oxide thickness (EOT) ≤ 0.5 nm. MOS structure with high-k has been simulated in Technology Computer-Aided Design (T-CAD) to consider as gate stack: metal/oxide/p-Si for the different FAB nodes; 45nm, 32nm, 22nm, and 14 nm.
- Homogeneous Field-Effect Transistors (III-V Submicron Metal Semiconductor Field Effect Transistor—MESFET) Homogeneous Field-Effect Transistors (III-V Submicron MESFET and Junction Field Effect Transistor—JFET).
- Heterostructure Field-Effect Transistors (III-V Submicron Heterostructure Field Effect Transistor—HFET) Heterostructure Field-Effect Transistors (III-V Submicron High-Electron Mobility Transistor—HEMT).
- III-V Submicron Metal-Oxide Semiconductor Field-Effect Transistor—MOSFET and Gallium Nitride (GaN) HEMT.
- Bipolar Transistor Operation, Silicon Bipolar Transistor Heterojunction Bipolar Transistor (III-V Submicron Heterojunction Bipolar Transistor—HBT).
- Scaled MOSFETs, CMOS/Bi CMOS Strain-Si, and Silicon Germanium (SiGe)-based MOSFETs.
- Ge MOSFET and carbon nanotube field-effect transistor (CNTFET).
- Power MOSFET Si Laterally Diffused Metal-Oxide FET (LDMOSFET) semiconductor.
- Silicon Carbide (SiC) Power Devices Germanium Nitride (GaN) Power Devices.
- Quantum-Effect Devices (Resonant-Tunneling Diodes) Hot-Electron Devices.
- Single-Electron-Transistors (SET) and Quantum Dots.
- THz transistor—High-Speed Photonic Devices (LED, Pin Photodetector, Avalanche Photodetector).
- High-Speed Photonic Devices (Laser).

The 5G/6G key performance indicators (KPIs), such as area traffic capacity (e.g., and connection density: 10 Mbps/meter^2 for 5G and 1 Gbps/meter^2) and connectivity densities (that is, the number of subscribers/devices are connected in the 5G/6G cellular system simultaneously, demand high-speed semiconductor for the devices.

1.5 EMERGENCE OF EXTREMELY HIGH-BANDWIDTH NETWORKING TECHNOLOGIES

1.5.1 TERRESTRIAL WIRELINE NETWORKS

The long-haul high-speed terrestrial backbone networks have been built with the synchronous optical network (SONET)/time division multiplexing (TDM) network, and dense wavelength division multiplexing (DWDM)/optical fiber. In addition, Wi-Fi, time division multiple access (TDMA), code division multiple access (CDMA), orthogonal frequency division multiple access (OFDMA), microwave access (WiMAX), and long-term evolution (LTE) have enabled the efficient multiplexing of high-speed users.

1.5.2 FIFTH- AND SIXTH-GENERATION WIRELESS NETWORKS

Fifth-generation (5G) technology enables a new kind of network that is designed to connect virtually everyone and everything together, including machines, objects, and devices. 5G wireless technology is meant to deliver higher multi-Gbps peak data speeds, ultra-low latency, more reliability, massive network capacity, increased availability, and a more uniform user experience to more users. Higher performance and improved efficiency empower new user experiences and connect new industries. Sixth-generation (6G) technology will be classified as the sixth-generation mobile network and the future successor of 5G [2]. The 6G network will be able to use higher frequencies than its predecessor, 5G, with improved capabilities of speed, high-level data processing, seamless connectivity, and significantly reduced latency.

1.5.3 HIGH-BANDWIDTH SATELLITE NETWORKS

The satellite networks will play a role in 6G by complementing the terrestrial network for coverage extension, especially in remote and rural areas (e.g., deserts, mountains, forests), where it is not economically viable to deploy the fixed infrastructures and for in-flight communications. Recent advances in reusable rocket technologies, introduction of low-cost satellites, smaller satellites (e.g., cube satellites at Low Earth Orbits) [2], and increasing demand for in-flight internet access, vessel tracking, and natural disasters monitoring have already set the stage for using non-terrestrial network (NTN) to complement terrestrial network (TN) connectivity. Third Generation Partnership Project (3GPP) Release-17 will support 5G new radio (NR) and long-term evolution internet-of-things (LTE-IoT) using NTN. Hence, there is an impending need to design system aspects and solutions for supporting NTN for ubiquitous network coverage and device connectivity in 6G.

NTN raises some new challenges in system and network architecture. This includes large round-trip time (RTT), high differential delay (RTT difference between cell-edge and cell-center UEs), and frequent handover and cell reselection, arising from high mobility in non-geostationary satellites. Note that AI, machine learning, and deep learning (AI/ML/DL) could enhance network operation by tuning available parameters and procedures for the best fit to traffic. For instance, context awareness may be used to predict the best-fitting configuration of lower layers in a radio node or to provide service proactively based on network-provided models and/or user-specific behaviors.

1.6 OPEN SYSTEMS INTERCONNECTION

The open systems interconnection (OSI) model [3] is a conceptual model created by the International Organization for Standardization, which enables diverse communication systems to communicate using standard protocols. The layers are Layer 1-Physical, Layer 2-Data Link, Layer 3-Network, Layer 4-Transport, Layer 5-Session, Layer 6-Presentation, and Layer 7-Application (Figure 1.1). The OSI model describes seven layers that computer systems use to communicate over a network.

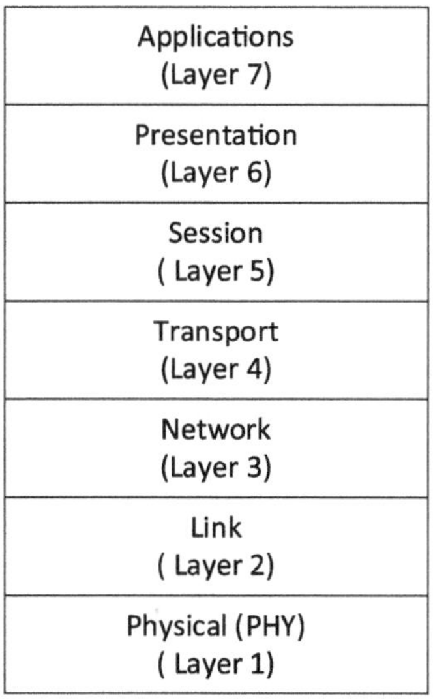

Applications (Layer 7)
Presentation (Layer 6)
Session (Layer 5)
Transport (Layer 4)
Network (Layer 3)
Link (Layer 2)
Physical (PHY) (Layer 1)

FIGURE 1.1   OSI Protocol Layer.

It was the first standard model for network communications, adopted by all major computer and telecommunication companies in the early 1980s. A full treatment of OSI model is provided in the book (see Chapter 14).

The model guides the product manufacturers how to build hardware and software to build their networks providing interoperability in multi-vendor environments. Networked AI means all these protocol layers will be AI-enabled. This book focuses on how communications network protocols will be AI-enabled in offering networking services.

1.7   INTERNET PROTOCOL LAYER MODEL

The modern Internet is based on the simpler transmission control protocol (TCP)/user datagram (UDP) in layer 4 and internet protocol (IP) in layer 3 model (Figure 1.2). Layers 5 and 6 are defined as the middleware layer that is used by the application layer for creation of services. However, layers 1 and 2 are the same as defined in the OSI model.

In this book, we will be describing how each of these internet protocol layers will be made AI-enabled for providing networking services. In addition, we have described this protocol model in detail in Chapter 14.

1.8   SUMMARY

In this subsection, we have formally introduced the networked AI architecture that provides the framework of communications and computation by AI. In addition, we have described machine learning, deep learning, and neural networks. We have described briefly AI-enabling and supporting technologies such as NLP, computer vision, graphical processing units, internet-of-things, advanced algorithms, and APIs, as well as emergence of extremely high-speed semiconductors, extremely high-bandwidth networking such as terrestrial and non-terrestrial technologies. We have discussed standardized communications network layer protocol model of both OSI and Internet protocol later that will be used to define the networked AI for both communications and computing in the subsequent sections.

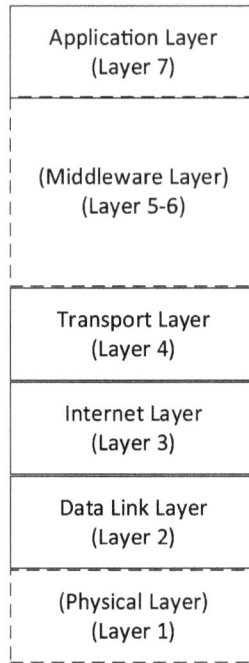

FIGURE 1.2   Internet Protocol Layer Model.

REFERENCES

[1] IBM. https://www.ibm.com/topics/artificial-intelligence.
[2] Next G Alliance Report: 6G Technologies, June 2022.
[3] Zimmermann, H., "OSI Reference Model—The ISO Model of Architecture for Open Systems Interconnection," IEEE Transactions on Communications. 28 (4): 425–432, April 1980.

2 Artificial Intelligent Agent

In AI, an agent is a computer program or system that is designed to perceive its environment, make decisions, and take actions to achieve a specific goal or set of goals. The agent operates autonomously, meaning it is not directly controlled by a human operator.

2.1 AGENT AND ENVIRONMENT

An environment is everything in the world which surrounds the agent, but it is not a part of an agent itself. An environment can be described as a situation in which an agent is present. The environment is where an agent lives, operates, and provides the agent with something to sense and act upon it. The relationship between the agent and environment is depicted in Figure 2.1.

The agent acts on the environment, and the environment sends back stimuli (that is, physical or behavioral changes or past experiences) to the agent triggered by the action of the agent. If an agent sensor can sense or access the complete state of an environment at each point of time, then it is a fully observable environment; else it is partially observable. A fully observable environment is easy as there is no need to maintain the internal state to keep track of the history of the world. If an agent has no sensors in all environments, then such an environment is called unobservable. An agent might have external inputs or knowledge such as abilities, goals/preferences, and prior knowledge (Figure 2.2).

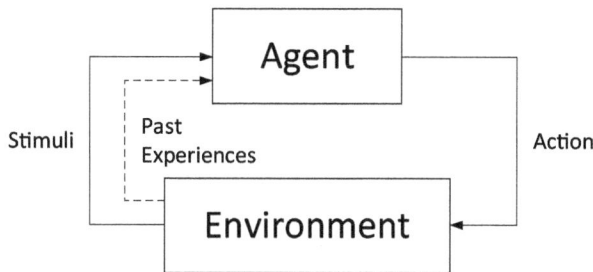

FIGURE 2.1 Agent Interacting with Environment.

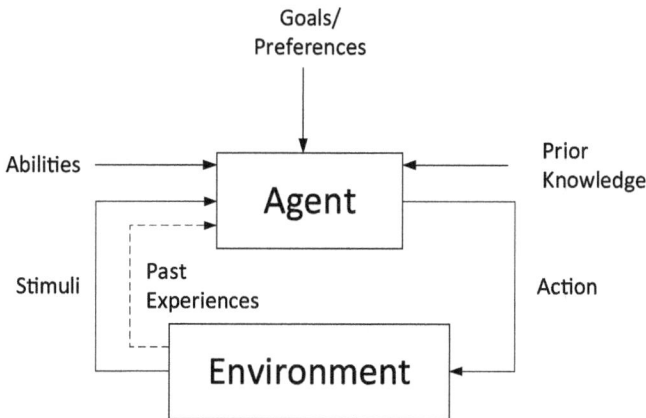

FIGURE 2.2 An Agent with External Inputs.

DOI: 10.1201/9781003499466-2

Abilities mean the primitive actions an agent can carry out. Goals mean that the agent must try to achieve or** preferences over states of the world. Prior knowledge is about the agent and the environment. Note that all the inputs that the agent is perceiving now as stimuli, abilities, goals/preferences, prior knowledge, and past experiences are termed as the percept.

2.2   AGENT SYSTEM

An agent's system consists of the agent and the environment in which the agent acts. The agent receives stimuli from the environment and carries out actions in the environment. Again, the agent itself is made up of a body and a controller as shown in Figure 2.3. The controller receives percepts from the body and sends commands to the body.

2.2.1   CONTROLLER

The controller receives real-time performance data from agents, visualize the application performance, and sends instructions to the agent (Figure 2.3). A controller maintains the agent's belief state and determines what command to issue at each time. The information it has available when it must do this is its belief state and its current percepts. A control problem is separable if the best action can be obtained by first finding the best model of the world and then using that model to determine the best action.

2.2.2   HIERARCHICAL CONTROLLER

Hierarchical control systems use a hierarchy of controllers (Figure 2.4). Each layer sees the layers below it as a virtual body from which it gets percepts and to which it sends commands.

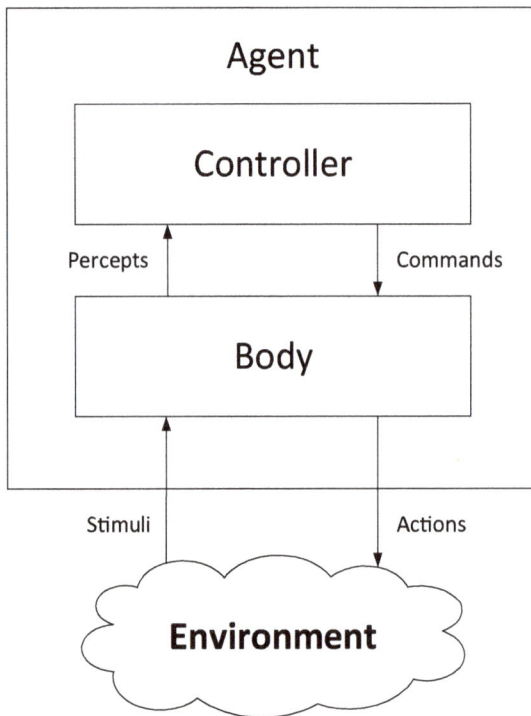

FIGURE 2.3   Agent System.

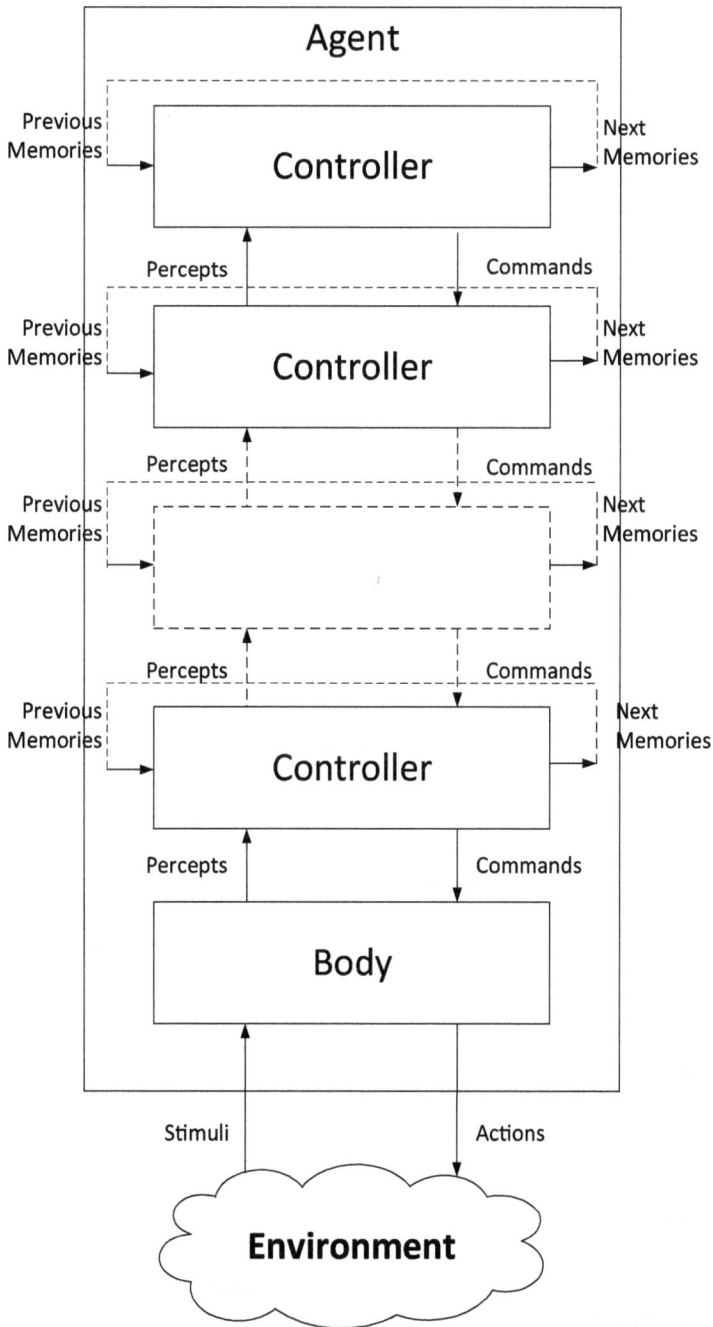

FIGURE 2.4 Idealized Agent System with Hierarchical Controller.

The lower-level layers run much faster, react to those aspects of the world that need to be reacted to quickly, and deliver a simpler view of the world to the higher layers, hiding details that are not essential for the higher layers [1]. The planning horizon at lower levels is typically much shorter than the planning horizon at upper levels. People must react to the world, at the lowest level, in fractions of a second but plan at the highest level even for decades into the future. For example, the reason for doing some tasks in an agency may be for the long-term goal, but reading and answering a question of a customer in a retail store must happen in minutes, if not seconds.

In another example, the low-level controllers might be operating small parts of the plant while higher-level controllers are controlling the actions of subordinate controllers. Using a hierarchical structure establishes clear authority for work and departments. Managers have authority according to management level and have the power to allocate resources, reward and punish behavior, and give orders to their subordinates. There are three types of inputs to each layer at each time: [1]

- Features that come from the belief state stored in memories (Figure 2.4), which are referred to as the remembered or previous values of these features
- Features representing the percepts from the layer below in the hierarchy
- Features representing the commands from the layer above in the hierarchy

Thus, a layer implements the following:

$$remember : s \times p_l \times c_h \rightarrow s$$

$$Command : s \times p_l \times c_h \rightarrow c_l$$

$$tell : s \times p_l \times c_h \rightarrow p_h$$

where s is the belief state, C_h is a set of commands from the higher layer, P_l is the set of precepts from the lower layer, C_l is the set of commands, and P_h is a set of precepts for the higher layer. Computing these functions can involve arbitrary computation, but the goal is to keep each layer as simple as possible. To implement a controller, each input to a layer must get its value from somewhere. Each percept or command input should be connected to an output of some other layer. Other inputs come from the remembered beliefs (stored in memories—see Figure 2.4). The outputs of a layer do not have to be connected to anything, or they could be connected to multiple inputs.

2.2.3 BODY

A body includes sensors that convert stimuli into percepts and actuators, also called effectors, which convert commands into actions. A body can also carry out actions that don't go through the controller, such as a stop button for a robot and reflexes of humans.

2.2.4 ENVIRONMENT

An environment is everything in the earth surrounding the agent, but it is not a part of an agent itself. An environment can be referred to as a situation in which an agent is present. The environment is where the agent lives, works, and provides the agent with something to sense and act upon it. The environment is mostly described as non-feministic. An environment might have a variety of characteristics from the perspective of an agent: [2] Fully observable vs Partially Observable, Static vs Dynamic, Discrete vs Continuous, Deterministic vs Stochastic, Single-Agent vs Multi-Agent, Episodic vs Sequential, Known vs Unknown, and Accessible vs Inaccessible [2].

- **Fully Observable vs Partially Observable**: A fully observable environment is one in which an agent sensor may perceive or access the entire state of an environment at any given time; otherwise, it is partially observable. It's simple to create a completely observable environment because there's no need to keep track of the world's past. Unobservable environments are those in which an agent has no sensors in all environments.

- **Deterministic vs Stochastic**: If an agent's current state and selected action can perfectly predict the forthcoming state of the environment, then such an environment is a deterministic environment. An agent cannot entirely control a stochastic environment because it is unpredictable in nature. Agents do not need to be concerned about uncertainty in a deterministic, fully observable world.
- **Episodic vs Sequential**: In an episodic environment, there is a succession of one-shot actions that just require the present percept. In a Sequential context, however, an agent must remember previous acts to decide the next best action.
- **Single-Agent vs Multi-Agent**: The term "single agent environment" refers to an environment in which just one agent is present and operates independently. A multi-agent environment, on the other hand, is one in which numerous agents are functioning in the same space. The issues of agent design in a multi-agent environment differ from those in a single-agent environment.
- **Static vs Dynamic**: If the environment can vary while an agent is deliberating, it is referred to as a dynamic environment; otherwise, it is referred to as a static environment. Static settings are simple to deal with because an agent does not need to keep glancing around while deciding. Agents, on the other hand, in a dynamic environment must continuously glance about at each action.
- **Discrete vs Continuous**: A discrete environment is one in which there are a finite number of percepts and actions that can be performed within it, whereas a continuous environment is one in which there are an infinite number of percepts and actions that may be performed within it. A chess game takes place in a discrete context since there are only so many moves that may be made. A continuous environment is exemplified by a self-driving automobile.
- **Known vs Unknown**: The terms "known" and "unknown" do not refer to features of the environment but rather to an agent's state of knowledge when performing an action. The effects of all actions are known to the agent in a known environment. To perform an action in an unfamiliar environment, the agent must first learn how it operates. A known environment could be partially observable, whereas an unknown environment could be entirely observable.
- **Accessible vs Inaccessible**: If an agent can acquire complete and correct knowledge about the state's environment, it is referred to as an accessible environment; otherwise, it is referred to as an inaccessible environment. An accessible environment is an empty room whose condition may be characterized by its temperature. An example of an inaccessible environment is information regarding a global incident.

2.3 AGENT ARCHITECTURE AND CONTROL

Agent architecture has been one of the core components in building an agent application as depicted in Figures 2.2 and 2.3. Agent architecture is considered as the functional brain of an agent in making decisions and reasoning to solve problems and achieving goals. In fact, agent architecture is a blueprint for software agents and intelligent control systems, depicting the arrangement of components. The architectures implemented by intelligent agents are referred to as cognitive architectures. The term agent is a conceptual idea but not defined precisely. It consists of facts, a set of goals, and sometimes a plan library.

There can be three main types of architectures in AI: symbolic, connectionist, and evolutionary. Each has its own benefits and drawbacks. Symbolic architectures are based on logic and reasoning. They can be very powerful, but they can also be inflexible. Connectionist architectures are based on neural networks. They are more flexible than symbolic architectures, but they can be less powerful.

Evolutionary architectures are based on evolutionary algorithms. They are very flexible and can be very powerful, but they can also be difficult to design. A key problem in such *architectures* is what kind of *control* framework to embed the *agent's* subsystems in to manage the interactions between layers.

2.4   SUMMARY

We have described the characteristics of AI agent, environment, agent system, controller and hierarchical controller, and body. Hierarchical controller percepts and commands are discussed. In addition, different kinds of environments are described in detail. In addition, different types of agent architectures and controls are explained.

REFERENCES

[1] Poole, D. L. and Mackworth, A. K., "Artificial Intelligence: Foundation of Computational Agents," Second Edition, 2017, Cambridge University Press.
[2] https://tutorialforbeginner.com/agent-environment-in-ai.

3 Agent Function

The agent function is a mathematical function that maps a sequence of perceptions into action. The function is implemented as the agent program. The part of the agent taking action is called an actuator. Agents are situated in time that can be discrete or dense. They receive sensory data in time and do actions in time. The action that an agent does at a particular time is a function of its inputs. Artificial intelligence agents perform these functions continuously: [1] perceiving dynamic conditions in the environment, acting to affect conditions in the environment, using reasoning to interpret perceptions, problem-solving, drawing inferences, and determining actions and their outcomes.

3.1 TIME

Discrete time views values of variables as occurring at distinct, separate "points in time" or equivalently as being unchanged throughout each non-zero region of time termed as time, that is, time is viewed as a discrete variable. Chess is classified as a discrete time AI problem. In contrast, continuous time views variables have a particular value only for an infinitesimally short amount of time. Between any two points in time there are an infinite number of other points in time. The variable "time" ranges over the entire real number line or depending on the context, over some subset of it such as the non-negative reals. Thus, time is viewed as a continuous variable. Continuous time AI environments rely on unknown and rapidly changing data sources. Multi-player video games are a classic example of continuous time AI environments.

3.2 TRACING PERCEPT

A percept trace is a sequence of all past, present, and future percepts received by the controller.

3.3 TRACING COMMAND

A command trace is a sequence of all past, present, and future commands output by the controller.

3.4 HISTORY

An agent's history at time t is a sequence of past and present percepts and past command. An agent doesn't have access to its entire history. It has access to only what it has to remember.

3.5 MEMORY/BELIEF STATE

The memory or belief state of an agent at time t encodes all the agent's history that it has access to. The belief state of an agent encapsulates the information about its past that it can use for current and future actions. The memory or belief state of an agent at time t encodes all the agent's history that it has access to. The belief state of an agent encapsulates the information about its past that it can use for current and future actions. A knowledge base (KB) might be created to store all belief states in addition to capturing human expert knowledge to support decision-making, problem-solving, and more. Through the years, knowledge base systems have been developed to support many organizational processes. One of the most transformational applications has been AI-powered customer support, including self-service options.

DOI: 10.1201/9781003499466-3

3.6 RATIONALITY

Perfect rationality refers to the ability to generate or choose behavior that will bring maximum success, given the situation and available information. It constrains the ability of an agent to provide the maximum expectation of success while considering the information that is available. Bounded rationality describes the way humans make decisions that depart from perfect economic rationality since we are limited by our mental capacity, the information available to us, and time. Instead of striving to make the "best" choices, we often settle on making merely satisfactory choices.

3.7 ONLINE AND OFFLINE LEARNING

Online learning is more likely to be the best option where machine learning engineers deal with large amounts of data, such as in areas like finance and economics, where new data patterns emerge on an almost constant basis. Offline or batch learning is more likely to be the best option where data patterns are consistent, emerge more slowly, and where there are no rapid concept shifts.

3.8 ACTION

Agent is a part of *AI* system that takes *actions* or decisions based on the information it perceives from the environment. In AI, acting is the process of taking an action upon environment by an AI agent toward achieving goals based on reasoning after analysis of present observation, knowledge base (KB), prior and past experiences. AI agent needs to act humanly means that computer programs must behave according to certain normal conventions of human interaction to make themselves understood. For example, natural language dialogues need to be set up with users. Note that Turing test is not used in AI in the industry in designing an AI agent that acts like humans because of complexity and lack of sufficient clarity.

A big area has emerged when AI agent is designed to act rationally using some sort of cognitive models that mimic human thought processes (e.g., verbal, visualizing, memory use, verbal relations, perceptual speed, induction, and deduction), including machine learning, deep learning, neural networks, natural language processing (NLP), and sentiment analysis. For example, emulating cognitive skills by an agent needs to pass all Turing tests to prove that an agent is rational. An agent needs to have the ability to represent knowledge and reason with it because this enables it to reach good decisions in a wide variety of situations. For example, an agent can be developed with better natural language knowledge to create sentences or dialogues to help humans to get by in a complex society worldwide. Similarly, more in-depth better visual perception generated by an agent will allow us to get a better idea of what an action might achieve. An agent acting rationally is more amenable to scientific development because the standard of rationality is clearly defined and completely general.

3.8.1 SOLUTION SEARCHING

The solution to a search problem is a sequence of actions in search space/state space called the plan that transforms the start state to the goal state. This plan is achieved through search algorithms. Each point in the search space represents one possible solution. Each possible solution can be "marked" by its value (or fitness) for the problem. All search methods essentially fall into one of two categories: exhaustive (blind) methods and heuristic or informed methods.

3.8.2 REASONING

A rational *agent* is a computer program that uses *logical reasoning* and the ability to make decisions to determine its following *action*. It uses a set of rules to determine the best course of action for a given situation. These agents usually work by comparing their current state with their previous state and then choosing an action based on how much better or worse they feel about their position now compared to before. The most basic form of this type of agent is called a reinforcement learning agent, which gets its name from the fact that it learns from experience as it goes along—it tries things out and then rates them based on how well they worked out in the past.

3.8.3 PLANNING

An agent which responds to goals through forming plans to achieve them and then enacting these plans through interaction within a domain. The goal of *action planning* is to choose actions and order relations among these actions to achieve specified goals. In AI planning, planners typically input a domain model (a description of a set of possible actions which model the domain) as well as the specific problem to be solved specified by the initial state and goal in contrast to those in which there is no input domain specified.

- **Classical Planning**: Three tasks are performed in classical planning: planning, acting and learning. In planning, the agent plans after knowing what the problem is, in acting, it decides what action it must take. In learning, the actions taken by the agent make him learn new things.
- **Planning and Acting in Real World**: *Planners* in the *real world* require extensions, representation language, actions with duration and resource constraints.
- **Planning with Certainty:** If the initial state is known unambiguously in a deterministic domain, planning with certainty can be performed, and all actions will be deterministic, and the state of the world after any sequence of actions can be accurately predicted.
- **Planning with Uncertainty**: Planning with uncertainty is performed for environments where there may be incomplete or faulty information, where actions may not always have the same results and where there may be tradeoffs between the different possible outcomes of a plan.
- **Relational Planning:** Relational planning provides useful state abstractions that enable faster learning and efficient transfer across tasks. The benefit of using the state abstractions is critical in relational settings, where the number and/or types of objects are not fixed a priori. Relational planning framework not only achieves better performance and efficient learning on the task at hand but also demonstrates better generalization to unseen tasks.

3.8.4 ACTION WITH CERTAINTY

Certainty is the epistemic property of beliefs which a person has no rational grounds for doubting. It is also known as epistemic certainty or objective certainty. There are three different types of certainty: opinion, belief, and conviction. In deterministic domains, the outcomes of action executions can be predicted with certainty. Many traditional domains from AI are deterministic, including sliding-tile puzzles and blocks worlds. Agent-centered search methods can solve offline planning tasks in these domains by moving a fictitious agent in the state space. In this case, the local search spaces are not imposed by information limitations. Agent-centered search methods thus provide alternatives to traditional search methods. They have, for example, successfully been applied to optimization and constraint-satisfaction problems and are often combined with random restarts. Examples include hill climbing, simulated annealing, tabu search, and some scheduling methods. Agent-centered search methods have also been applied to traditional search problems and planning problems. For these planning problems, agent-centered search methods compete with other heuristic search methods such as greedy (best-first).

3.8.5 Action with Uncertainty

Dealing with uncertainty is also an important issue in the field of AI: AI systems should know when they do not know something, when they should be cautious, and when they need to take risks, especially in high-risk application areas, such as critical infrastructure, health care, or the military. Risk and ambiguity are the two fundamental forms of uncertainty in decision-making. **Risk** is present whenever, in a known situation, the outcome of an event is uncertain, but probabilities can be computed, as in dice, for example. **Ambiguity**, on the other hand, refers to unknown situations in which the probabilities are not known or cannot be determined. Many robotics, control, and scheduling domains are nondeterministic.

Planning in nondeterministic domains is often more difficult than planning in deterministic domains because their information limitation can be overcome only by enumerating all possible contingencies, resulting in large search spaces. Consequently, it is even more important that agents take their planning cost into account to solve planning tasks efficiently. Agent-center research in nondeterministic domains has an additional advantage over agent-centered search in deterministic domains, namely that it allows agents to gather information early. This advantage is an enormous strength of agent-centered search because this information can be used to resolve some of the uncertainty and, thus, reduce the amount of planning performed for unencountered situations.

Without interleaving planning and plan execution, an agent must determine a complete conditional plan that solves the planning task, no matter which contingencies arise during its execution. Such a plan can be large. When interleaving planning and plan execution, however, the agent does not need to plan for every possible contingency. It must determine only the beginning of a complete plan. After the execution of this subplan, it can observe the resulting state and then repeat the process from the state that resulted from the execution of the subplan instead of all states that could have resulted from its execution.

Of course, two kinds of uncertainty can be present at the same time: the agent can be uncertain about how uncertain the outcome of an action is. Crucially though, if we only look at the predictive probabilities, the two kinds of uncertainty cannot be disentangled: being uncertain about certain outcomes and being certain about uncertain outcomes may result in the same predictions as to what might happen. The only thing we care about is predictive probabilities, therefore, we disentangle model uncertainty from the environment. This is because in some cases like classification it may not matter that much, but in certain cases, it may make or break the method.

3.8.6 Action in Learning

Action in learning is a process of insightful questioning and reflective listening. Action learning tackles problems through a process of first asking questions to clarify the exact nature of the problem, reflecting, and identifying possible solutions, and only then taking action. Business simulations are an example of the action learning approach. The participants take on the management role of a business and learn to make management decisions in a gamified way under time pressure. There are many benefits of action model learning in AI. One benefit is that it can help agents learn how to perform tasks more efficiently by observing how other agents perform the same tasks. Additionally, action model learning can help agents learn how to generalize their knowledge to new situations and domains.

3.9 SUMMARY

We have described AI agent functions in detail, including time, tracing precept, tracing command, history, memory/belief state, rationality, offline and online learning, and action. We have explained each function of action of the agent: solution searching, reasoning, action with certainty, action with uncertainty, and action in learning.

REFERENCE

[1] Poole, D. L. and Mackworth, A. K., "Artificial Intelligence: Foundation of Computational Agents," Second Edition, 2017, Cambridge University Press.

4 Agent Modeling

4.1 AGENT

In the field of artificial intelligence (AI), a broad consensus is that an agent is defined as a functional entity that processes the input information perceived from its environment and acts upon that environment [1]. In addition, the simplest artificial intelligent agent consists of a knowledge base (KB) and an inference engine as conceptualized in Figure 4.1. The agent together with its environment and/or other external entities influencing the agent is called the world. The input information that is received by the agent from the environment for processing is usually in the form of observations (e.g., measurement data) of the environment (e.g., sensors), instructions (e.g., goals and/or utility functions), and queries working in a loop.

In each iteration, the agent receives input in the form of observations of the environment, instructions, and queries. The inference engine carries out reasoning, which is driven by the data structures in the knowledge base (KB) to answer the queries and to choose actions to execute in the environment for achieving a goal. An agent may not achieve its goal with one iteration or one step. A plan is required what the agent will do at any given time and/or in the future. So planning is the process of finding a sequence of actions to achieve a goal by the agent.

A learner or machine learning algorithm is the program used to learn a machine learning model from data. Another name is "inducer" (e.g., "tree inducer"). A machine/deep learning model is the learned program that maps inputs to predictions. This can be a set of weights for a linear model or for a neural network. Perception in artificial intelligence is the process of interpreting vision, sounds, smell, and touch. Perception is a process to interpret, acquire, select, and then organize the sensory information from the physical world to make actions like humans.

4.2 REASONING

Reasoning is a broad area that could include human-like thinking even with rationality [1]. It is important to note that human beings have been taken as an ideal reference model to be emulated for building an AI agent, but in reality, no one is saying it is possible to reach that ideal goal ever. Human-like thinking requires the computer program expresses reasoning of human mind with sufficient precision. The cognitive theory from psychology that itself is a huge area has appeared to program the intelligent agent for thinking as close to human although its successful performance in this area is too far away in the distant future. Human-like thinking with rationality by an intelligent agent might include patterns for argument structures that yield correct conclusions. Most of the problems in artificial intelligence are rational reasoning problems. The argument structures which are rational to have correct answers or conclusions are stored in a knowledge base. Knowledge bases may use sematic search, natural language processing, symbolic reasoning, and other technologies for discovery of knowledge quickly and easily. Learning or knowledge engineering modifies the knowledge base with new experiences as more knowledge is acquired.

Knowledge base generally contains ontologies which define kinds of object that exist in the world and the attributes they possess, relationships that define how objects and attributes are related, goals and utilities which define desirable and undesirable states of the world and states of the agent as well as policies, control rules, and heuristics which define ways of acting. It is seen that the knowledge base may contain manually engineered (or automatically learned) policies that cannot be inferred (or explained) by the declarative knowledge in the knowledge base. Similar is the case for control rules and heuristics. Here comes machine and deep learning which can do better for learning policies, control rules, and heuristics with the knowledge that constraint and guide the inference

DOI: 10.1201/9781003499466-4

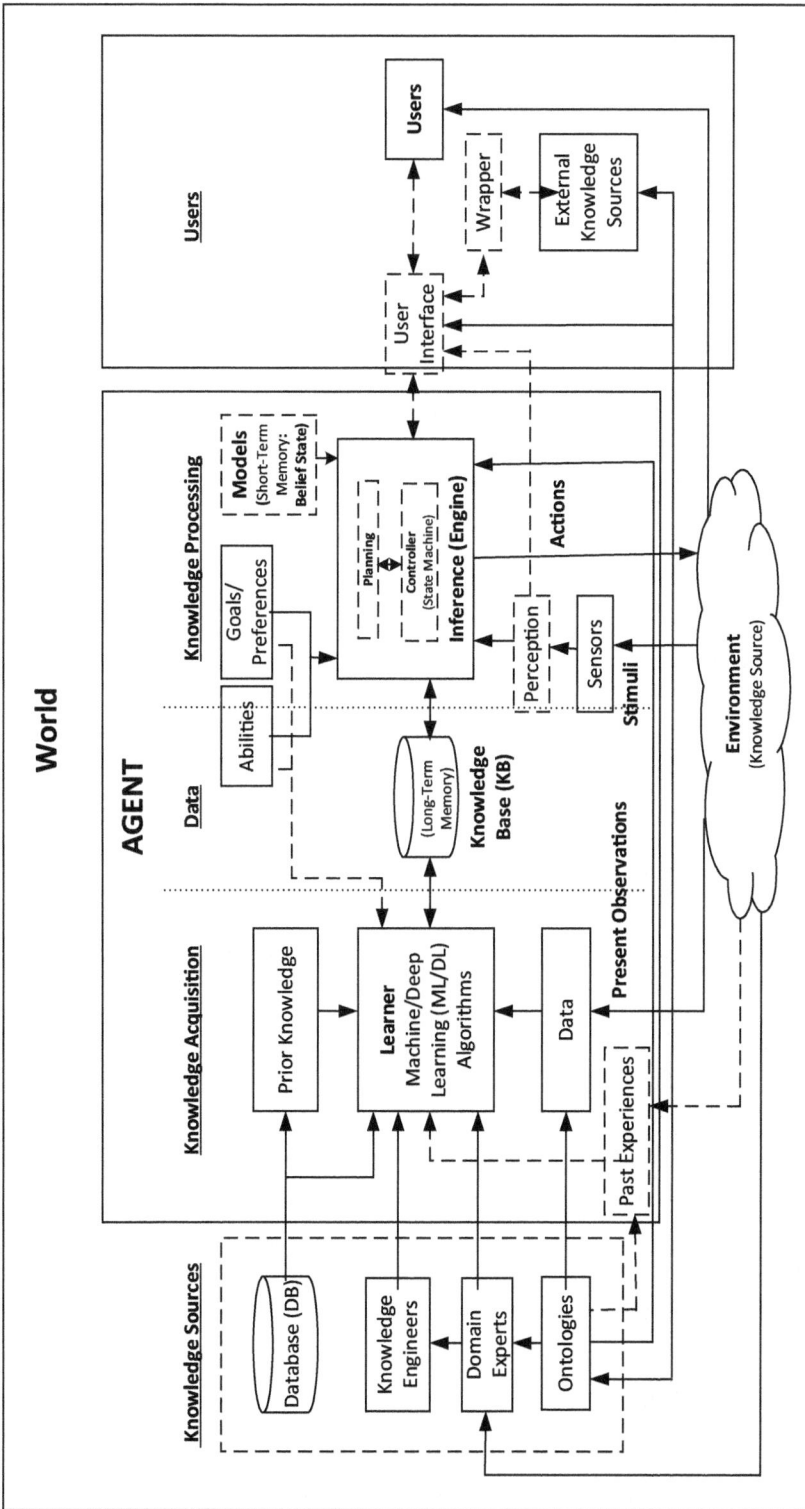

FIGURE 4.1 Conceptual Representation of Agent and World.

engine so that it arrives at the correct conclusion more efficiently meeting performance objectives. The key is that a plan that describes a sequence of actions to achieve a goal is the result of reasoning.

At present, machine learning has been developed completely decoupling with the declarative knowledge of reasoning avoiding the need for an inference engine and hence it has avoided the computational cost of reasoning. We have discussed machine learning separately in detail. Most of the problems in artificial intelligence are reasoning problems, and indeed both planning and CSP are among them. Planning is a subfield of AI devoted to finding action sequences that achieve an agent's goal.

4.2.1 COGNITIVE REASONING

Cognitive computing focuses on mimicking human behavior and reasoning to solve complex problems. AI augments human thinking to solve complex problems. It focuses on providing accurate results. It simulates human thought processes to find solutions to complex problems. Therefore, cognitive AI technology goes beyond conventional AI by utilizing a unique hybrid combination of machine/deep learning numeric approaches alongside higher-order symbolic techniques that deliver cognitive reasoning and intelligence resembling that of human intuition. One well-known example is AlphaGo, the AI that has bested multiple champions at the board game Go.

4.2.2 REASONING WITH CONSTRAINTS

Instead of reasoning explicitly in terms of states, it is typically better to describe states in terms of features and to reason in terms of these features, where a feature is a function on states. Features are described using variables. Often, features are not independent, and there are hard constraints that specify legal combinations of assignments of values to variables. The three primary constraints that project managers should be familiar with are time, scope, and cost. These are frequently referred to as the triple constraints or the project management triangle.

4.2.3 REASONING WITH UNCERTAINTY

Reasoning with uncertainty and time is the ability to draw logical conclusions and make decisions based on incomplete, inconsistent, or uncertain information that may change over time. Probabilistic reasoning is another name for reasoning under uncertainty. In real-world situations, agents are inevitably forced to make decisions based on incomplete data. Even when an agent senses the world to learn more, it rarely learns the precise state of the world. Uncertainty arises due to various factors such as unreliable sources of information, experimental errors, equipment faults, temperature variations, and climate change, among others.

4.2.4 KNOWLEDGE REASONING

Knowledge representation and reasoning is the field of AI dedicated to representing information about the world in a form that a computer system can use to solve complex tasks such as diagnosing a medical condition or having a dialog in a natural language. More precisely, the knowledge reasoning includes the following:

- **Propositions and Inference**: A propositional logic-based agent functions by expressing its understanding of the outside world as logical statements. The knowledge base is initially empty, but as the agent explores the environment, it fills it with fresh data. The agent draws new knowledge from its knowledge base through logical inference. For example, if we know that $P \wedge Q$ is true, and we also know that $P \rightarrow R$, we can infer that $Q \vee R$ is also true.

- **Individual and Relations Reasoning**: The representation dimension that concerns how the world is described has, at its top level, reasoning in terms of individuals entities and relations. Reasoning in terms of relations allows for compact representations that not only can be built independently of the particular entities but can also be used to learn and reason about each entity.
- **Ontologies**: An ontology is a basic term of knowledge as a collection of ideas within an area and their connections. Classes, individuals, characteristics, and relations, as well as rules, limitations, and axioms, must all be explicitly specified for such a description to be possible.
- **Knowledge-based Systems**: A knowledge-based system uses AI techniques to store and reason with knowledge. The knowledge is typically represented in the form of rules or facts, which can be used to draw conclusions or make decisions.
- **Probabilistic Reasoning**: Probabilistic reasoning is a form of knowledge representation in which the concept of probability is used to indicate the degree of uncertainty in knowledge. In AI, probabilistic models are used to examine data using statistical codes.
- **Knowledge Representation**: Knowledge representation and reasoning is the part of AI which concerned with AI agents thinking and how thinking contributes to intelligent behavior of agents. **Knowledge-based agents** have explicit representation of knowledge that can be reasoned. They maintain internal state of knowledge, reason over it, update it and perform actions accordingly. These agents act intelligently according to requirements.
- **Probabilistic Reasoning over Time**: It is the same as probabilistic reasoning, except that it includes time slice within which probabilistic reasoning is made. At best, the agent will be able to obtain only a probabilistic assessment of the current situation. An agent maintains a belief state that represents which states of the world are currently possible. From the belief state and a transition model, the agent can predict how the world might evolve in the next time-step. From the percepts observed and a sensor model, the agent can update the belief state.
- **Decisions Making**: Intelligent agents make decisions based on their perception of the environment and predefined goals.
- **Communicating and Perceiving**: In AI, an agent is a computer program or system that is designed to perceive its environment, make decisions, and take actions to achieve a specific goal or set of goals, and communicates with its partner(s). The agent operates autonomously, meaning it is not directly controlled by a human operator.

4.3   LEARNING

We have seen that reasoning requires knowledge, and knowledge is acquired through learning. Conceptually, learning is a process that improves the knowledge of an AI program by making observations about its environment. An agent uses prior knowledge and past experiences which is called learning to build knowledge base that is used for acting. Learning can use the online and/or offline training process for acquiring knowledge, experience, and behavior using computer programs based on observations about its environment for performing human-like tasks by an intelligent agent. Similar to humans, an AI agent will also have the ability to improve its past or prior knowledge, experience, and behavior based on ongoing new observations about its environment.

Perception uses sensory information while reasoning works mostly symbolically. These two are often joined by humans subconsciously. AI also exhibits two capabilities. Deep learning has successfully performed this task in solving perception tasks. At present, learning can be grouped into categories: learning with declarative knowledge of reasoning and learning with non-declarative knowledge of reasoning.

4.4   LEARNING WITH REASONING

Learning itself is a cognitive phenomenon while reasoning is also a major component of cognition. In learning and reasoning, both phenomena need to be present. So it implies both learning and reasoning need to be addressed under a common framework. Common semantics creation is the major challenge. Human problem-solving processes require that both perception and reasoning are two intelligent representations need integrated seamlessly. Perception and reasoning are two representative abilities of intelligence that are integrated seamlessly during human problem-solving processes. In the area of AI, the two abilities are usually realized by machine learning and logic programming, respectively. In AI, these two capabilities are realized using machine and deep learning which is termed as abductive learning. Abductive learning helps to recognize numbers and resolves unknown mathematical operations simultaneously from images of simple handwritten equations. Moreover, the learned models can be generalized to longer equations and adapted to different tasks, which is beyond the capability of state-of-the-art deep learning models.

4.5   FEDERATED LEARNING

In a federated learning setup, a machine learning model is trained on multiple decentralized servers controlled by different organizations, each with its own local data. They communicate with a central orchestrator that aggregates the individual model updates and coordinates the training process. Similar to distributed machine learning, federated learning also trains the models independently. The only difference between distributed machine learning and federated learning is that in federated learning, each participant initializes the training independently as there is no other participant in the network. Federated learning also comes in three categories such as horizontal federated learning, vertical federated learning, and federated transfer learning.

4.6   ENSEMBLE LEARNING

Ensemble methods are techniques that create multiple models and then combine them to produce improved results. These methods in machine learning usually produce more accurate solutions than a single model would. In this ensemble technique, machine learning professionals use several models for making predictions about each data point. The predictions made by different models are taken as separate votes. Subsequently, the prediction made by most models is treated as the ultimate prediction. As compared to the single-modeled ML disciplines described earlier, ensemble learning approaches can be more expressive and have less bias and variance. The benefit of ensemble learning is that its cost increases only linearly with the number of models in the ensemble rather than exponentially as it would be the case with a more general model. Assuming that the prediction or classification models are independent—which is a rather strong assumption—ensemble will make more accurate predictions and classifications. There are many ways of creating ensembles, including bagging, boosting, and random forests.

4.7   MACHINE LEARNING WITH LOGICAL REASONING

The ML technique is often used to construct abstractions while reasoning techniques can be leveraged to verify and correct them in a corrective feedback loop. These iterative learning and reasoning algorithms enable us to solve classes of AI problems that otherwise seem very difficult or infeasible to solve. Deductive reasoning is a type of reasoning in AI that involves concluding a set of premises or assumptions using logical rules. In deductive reasoning, the conclusion must also be true if the premises are true. Deductive reasoning is often used in mathematics and formal logic. Learning is one of the fundamental building blocks of AI solutions. From a conceptual standpoint, learning is a process that improves the knowledge of an AI program by making observations about its environment.

Unsupervised machine learning (see Chapter 8), supervised machine learning (see Chapter 9), deep learning (see Chapter 10), and reinforcement machine learning (see Chapter 13) also belong to machine learning with logical reasoning. Examples of unsupervised machine learning algorithms are centroid-based K-means clustering algorithm that is used for customer segmentation, target marketing segment, data exploration, data preparation, and other purposes in real-world example. In addition, principal component analysis (PCA) and hierarchical clustering are other two well-known unsupervised learning algorithms. Examples of supervised machine learning are linear regression, neural network with multiple perception, linear regression, support vector machine, Naïve Bayes, decision tree, K-nearest neighbor algorithm, similarity learning, and others. Image recognition, fraud detection, spam filters, classification, and regression are examples of supervised learning. Machine learning algorithms that use neural networks are known as deep learning. A list of machine and deep learning (ML/DL) algorithms is provided here.

- Machine Learning
  - Linear/Logistics Regression
  - Linear Discrimination Analysis
  - Gradient Boosted Decision Tree (XG Boost)
  - Factorization Machines
  - K-means/Extraction Maximization Clustering
  - Random Forest
  - Naive Bayes
  - Principal Component analysis
  - Linear Learner
  - Support Vector Machine
  - Belief Network
  - Hidden Markov Model (HMM)
  - Profile Hidden Markov Model (PHMM)
  - Fuzzy System and
  - Others
- Deep Learning
  - Deep Supervised Learning
    - Convolution Neural Network (CNN)
    - Recurrent Neural Network (RNN)
      - Long-Term Short-Term Memory (LSTM)
      - Gated Recurrent Unit
  - Deep Unsupervised Learning
    - Autoencoder
      - Restricted Boltzmann Machine
    - Self-Organizing Map
  - Neural Turing Machine (NTM)

4.7.1 HYBRID LEARNING

Most learning algorithms used in machine/deep learning (ML/DL) are good at completing one task or working with one dataset. While helpful and infinitely better than doing it manually, these algorithms won't help you realize the full potential of AI across all the data. That's where hybrid machine learning (HML) comes in. Multiple simple algorithms work together to complement and augment each other. Together they can solve problems that alone they were not designed to solve.

4.8   SUMMARY

We have described the AI agent modeling that consists of the agent itself (agent and world), reasoning, and learning. Reasoning consists of cognitive reasoning, reasoning with constraints, reasoning with uncertainty, and knowledge reasoning. Learning can be of different kinds such as learning with reasoning, federated learning, ensemble learning, and machine learning and hybrid learning. Machine learning belongs to the learning with logical reasoning. Deep learning is also a part of machine learning which uses neural networks. Unsupervised learning, supervised learning, and reinforcement learning are also briefly discussed here. Example algorithms that are used by each kind of learning are also listed here.

REFERENCE

[1] Poole, D. L. and Mackworth, A. K., "Artificial Intelligence: Foundation of Computational Agents," Second Edition, 2017, Cambridge University Press.

5 Multi-Agent System

5.1 OVERVIEW

A multi-agent system is one that consists of several agents, which interact with one another. In the most general case, agents will be acting on behalf of users with different goals and motivations. To successfully interact, they will require the ability to cooperate, coordinate, and negotiate with each other, much as people do. Building agents, we address questions such as:

- How do you state your preferences to your agent?
- How can your agent compare different deals from different vendors?
- What if there are many different parameters?
- What algorithms can your agent use to negotiate with other agents (to make sure you get a good deal)?

In multi-agent systems, we address questions such as:

- How can cooperation emerge in societies of self-interested agents?
- What kinds of languages can agents use to communicate?
- How can self-interested agents recognize conflict? and
- How can multi-agent systems be interdisciplinary?

The field of multi-agent systems is influenced and inspired by many other fields: philosophy, logic, game theory, economics, social sciences, and ecology. This can be both a strength, infusing well-founded methodologies into the field, and a weakness, there are many different views as to what the field is about and how they can (nevertheless) reach an agreement. How autonomous agents can coordinate their activities to cooperatively achieve goals. Multi-agent systems (MAS) are systems of independent software elements that can be used to aid humans in the process of taking decisions. They have been postulated as a suitable framework to deal with the complexity of industrial asset fleets formed by heterogeneous assets. Multi-agent systems have been especially successful in aiding humans to take decisions in complex environments such as traffic management, industrial production, and other areas.

The history of multi-agent systems is intrinsically linked to our understanding of the meaning of the word "agent," the definition described earlier (see Chapters 2–4). Agents are autonomous, problem-solving, and goal-driven computational entities with other functional entities. Multi-agent systems remain one of the most prolific frameworks to manage continuous monitoring systems, and recently they have been postulated as a way of providing assets with a certain degree of agency. From this idea, collaborative prognostics is the proposed framework in which agents share information with each other to improve computation and failure predictions, thus optimizing predictive maintenance.

A multi-agent system is defined by its architecture [1,2] that determines the structure and topology of its agents. Multi-agent system architectures have been broadly classified into four types: centralized, hierarchical, heterarchical, and distributed (or peer-to-peer). In collaborative prognostics, where agents are often linked with individual assets, the optimal architecture will be determined by its influence on the overall cost and reliability of the system. Collaborative prognostics in large fleets of assets comprehends several cost factors: communication, computational, and maintenance. Traditionally, maintenance costs were considered cardinal. With the advent of industrial

DOI: 10.1201/9781003499466-5

internet-of-things (IIoT) technologies, communication and computational costs have become relevant due to the large amount of data processed and transmitted through the internet in continuously monitored fleets.

When applied to predictive maintenance, several of the canonical architectures of multi-agent systems require dramatically increasing the amount of processing and communication within the fleet, as real-time peer-to-peer communication and prognostics are supported. State-of-the-art prognostics use a plethora of machine learning algorithms, which are often computation- and data-intensive. Therefore, it becomes crucial to quantify how maintenance costs compare to other costs to assess the suitability of different MAS architectures.

Here, we compare several canonical multi-agent architectures for collaborative prognostics since different cost balances between communication, maintenance, and computation. Concretely, we study the effect of varying asset value and communication costs in the overall cost of the architecture, and we show that different architectures are optimal for different industrial scenarios.

Apart of the cost constraints explicitly dealt with in this chapter, the implementation of a multi-agent architecture may be limited by other constraints such as human resources or available capital. This is especially important in the case of subject matter experts (SME) or industries operating in a context of low financial liquidity. We do not deal with such managerial details, but they must nonetheless be considered beforehand by any asset manager wishing to implement the proposed system in practice.

5.2   COLLABORATIVE PROGNOSTICS

The concept of collaborative prognostics extends the concept of collaborative agents into the field of prognostics and health management. Collaborative agents share information with each other to jointly achieve a given objective. In collaborative prognostics, machines (through their agents) behave like social entities, communicating with one another and taking their own decisions. In its core, collaborative prognostics involves formation of clusters of similar machines and collaboration among machines within these clusters to improve failure prediction and predictive maintenance. This collaboration can either be in the form of exchanging model parameters or condition data.

5.3   AGENT TYPOLOGIES AND FAILURE MODES

The architectures reviewed in this chapter are formed by four elements: virtual assets, digital twins, mediator agents, and a social platform. The agent's failure modes have been restricted to affect their deliberative and communicative capabilities. The experiments are set up under the assumption that there will be no data loss upon agent failure due to the widespread nature of backup systems in industry.

5.4   VALUE ASSET

Virtual assets are the lowest-level agents employed in collaborative prognostics. The virtual assets' tasks are limited to standardizing the data coming from their corresponding physical assets and sending that data to upper layers in the architecture. It must be mentioned that because of the rather simple tasks that they perform, virtual assets fail to satisfy some widely accepted definitions of agents. However, they are critical for the functioning of the system and thus we include them in our analysis.

Virtual assets act as passive nodes of the architecture and have no deliberative capabilities. Their data is divided into three main components: a set of sensor-produced features, a set of timed failures or warnings, and a unique identifier. Virtual assets are formed by two building blocks: a standardizer, dedicated to standardizing the data coming from their assigned assets, and a communications manager, which controls the communications with the upper layers of the architecture.

Failure: Virtual asset's failure corresponds to the severance of communications between a deteriorating asset and the rest of the architecture and thus the halt of prognostics for this particular asset.

5.5 DIGITAL TWIN

Digital twins are smart agents with prognostics, communication, and data-preprocessing capabilities. When digital twins are employed, each physical asset in the industrial system is assigned its individual digital twin. Digital twins are composed of three building blocks: an analytics engine, a data repository, and a communications manager. The analytics engine computes prognostics and the maintenance policy, the data repository manages the data available to the twin, and the communications manager controls the communication between the digital twin and other elements of the architecture. This includes the capability of independently choosing other twins to collaborate with.

Failure: The failure of a digital twin implies (1) that its communication with other agents is severed, (2) that the system stops providing maintenance recommendations for the physical asset assigned to the faulty digital twin, and (3) that the digital twin cannot perform any computation.

5.6 MEDIATOR AGENTS

Mediator agents are intermediate agents able to perform prognostics and determine the maintenance policy for groups of assets. They are also able to receive data from the virtual assets and send data to upper layers of the architecture. Mediator agents can communicate with each other through the social platform. Mediator agents are composed of the same building blocks as digital twins. However, their analytic engine and communications manager do not give them the capacity of choosing which agents to communicate with, as their communications are managed by the social platform.

Failure: The failure of a mediator agent implies (1) that its communication with other agents is severed, (2) that the system stops providing maintenance recommendations for the physical assets assigned to the mediator agent and (3) that the mediator agent stops using any computing power.

5.7 SOCIAL PLATFORM

The social platform is the agent serving as a central node in the centralized, hierarchical and heterarchical architectures. The main task of the social platform is to run algorithms leveraging information originating from the whole fleet. These algorithms can be aimed at (1) forming clusters of collaborating assets, (2) retrieving and plotting enterprise-level information, or (3) calculating prognostics and making maintenance decisions. Note that each of these tasks is optional and depends on the architecture in which the social platform is embedded.

The social platform uses data received from agents in lower layers of the hierarchy to form clusters of collaborating assets. In the case of a hierarchical architecture, the social platform acts also as a communication channel between the lower agents of the architecture. The platform is formed by three building blocks: a data repository, containing clustering information and the results of the algorithms run in the platform, an analytics engine where algorithms are computed, and a communication manager that controls communication with lower-level agents.

Failure: The failure of the social platform implies the severance of all communications and all computations managed by it. Additionally, in the centralized architecture, failure of the social platform implies the halt of all maintenance recommendations.

5.8 MULTI-AGENT ARCHITECTURE

In this section, we describe the architectures analyzed in this chapter. These architectures have been chosen because of their prominence in industrial systems. Here, we describe them within the context of collaborative prognostics.

5.8.1 Centralized

The centralized architecture is the simplest case considered here. It consists of a social platform with full control over the decision-making of the system, and a set of virtual assets that limit themselves to sending data to the social platform. the social platform computes the clusters of similar assets and then uses the data from the assets belonging to these clusters to generate maintenance recommendations (see Figure 5.1).

A centralized architecture can technically be argued to not be a multi-agent architecture, as the only agent that really takes decisions and outputs predictions is the social platform. Nevertheless, we decide to test it against other architectures because of its importance and widespread use in industrial applications.

5.8.2 Hierarchical

A hierarchical architecture is defined as an architecture in which intermediate agents provide most of the decision-making in the system while lower-level agents are left to perform simpler tasks. In our case, these intermediate agents are mediator agents. Mediator agents are assigned groups of virtual assets for which they perform prognostics and schedule maintenance actions (see Figure 5.2).

The social platform is hierarchically superior to the mediator agents and in fact assigns them to groups of similar assets. The social platform can also create or delete mediator agents (since the number and membership of clusters may vary over time) and has full control of the communications of the system.

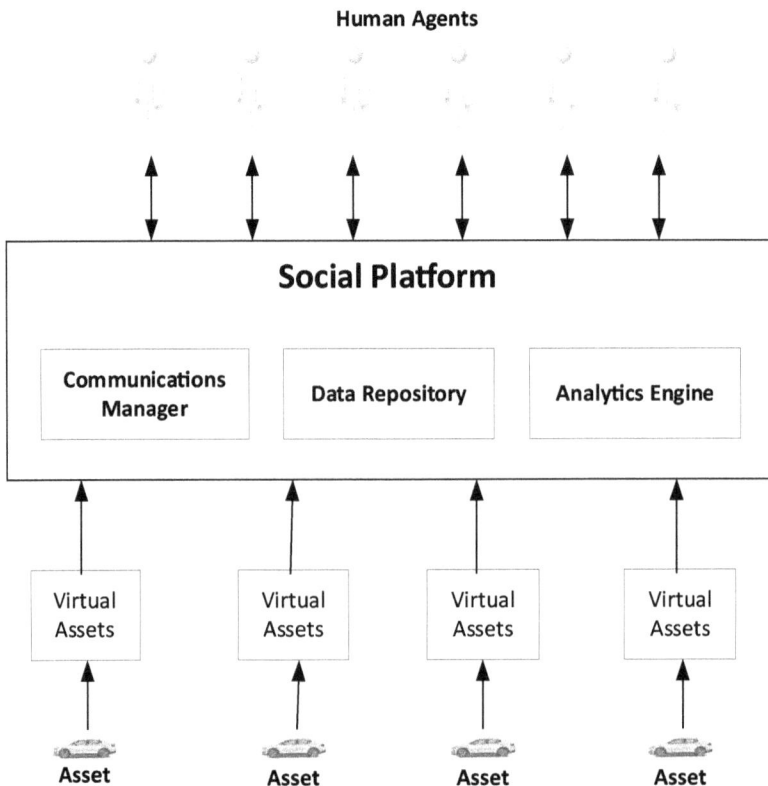

FIGURE 5.1 Centralized MAS Architecture.

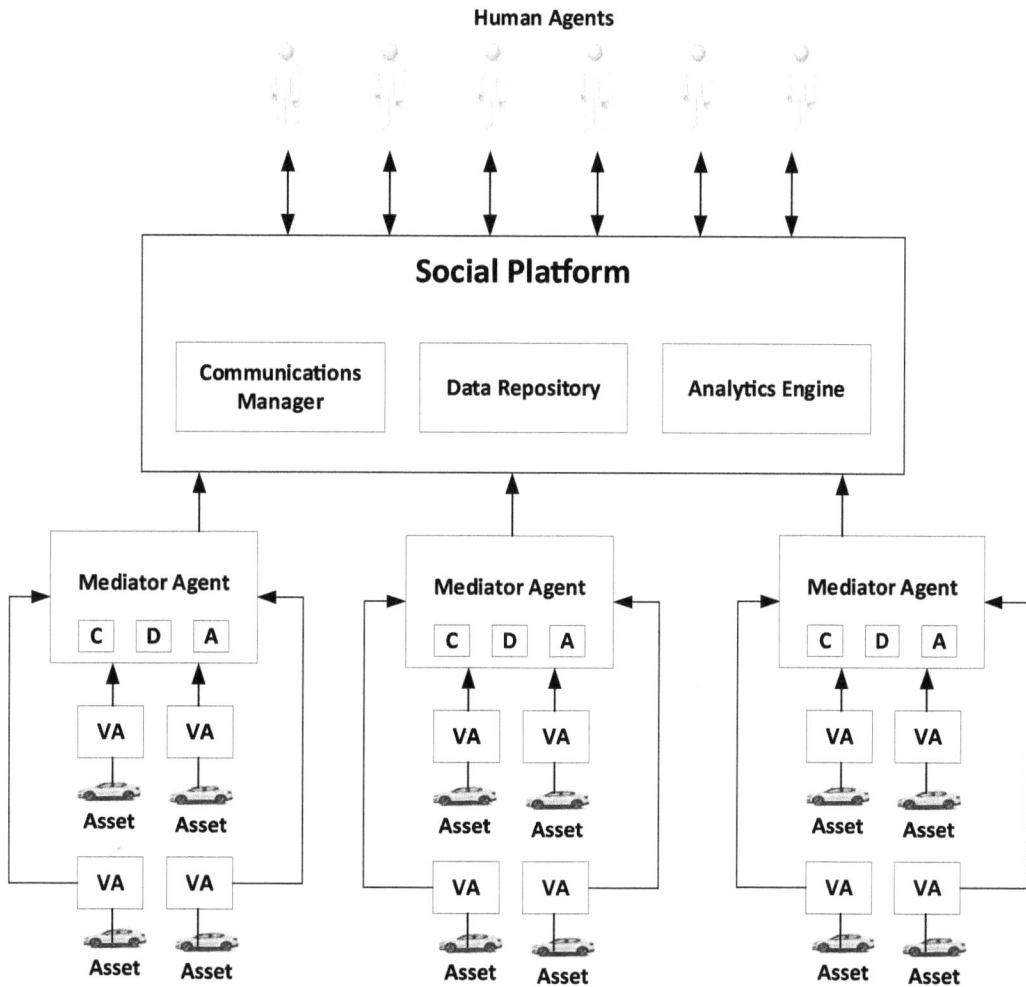

FIGURE 5.2 Hierarchical MAS Architecture.

5.8.3 HETERARCHICAL

A heterarchical architecture differs from the hierarchical case presented earlier in that it allows for peer-to-peer communication between the digital twins. Concretely, in our implementation of this architecture, digital twins perform prognostics, take maintenance decisions, and communicate with each other. The social platform, at a higher level in the architecture, decides which digital twins will communicate with each other through its clustering algorithm and serves as a communication link with human operators (see Figure 5.3).

5.8.4 DISTRIBUTED

A distributed architecture is one in which all its agents are in the same level of the hierarchy and could take independent decisions without the supervision of a higher-level agent (see Figure 5.4).

In this architecture, communication consists of peer-to-peer connections between digital twins (i.e., the twins are all connected to one another without any central agent or mediators present). As in the architectures described previously, similar assets are clustered together for collaborative

Human Agents

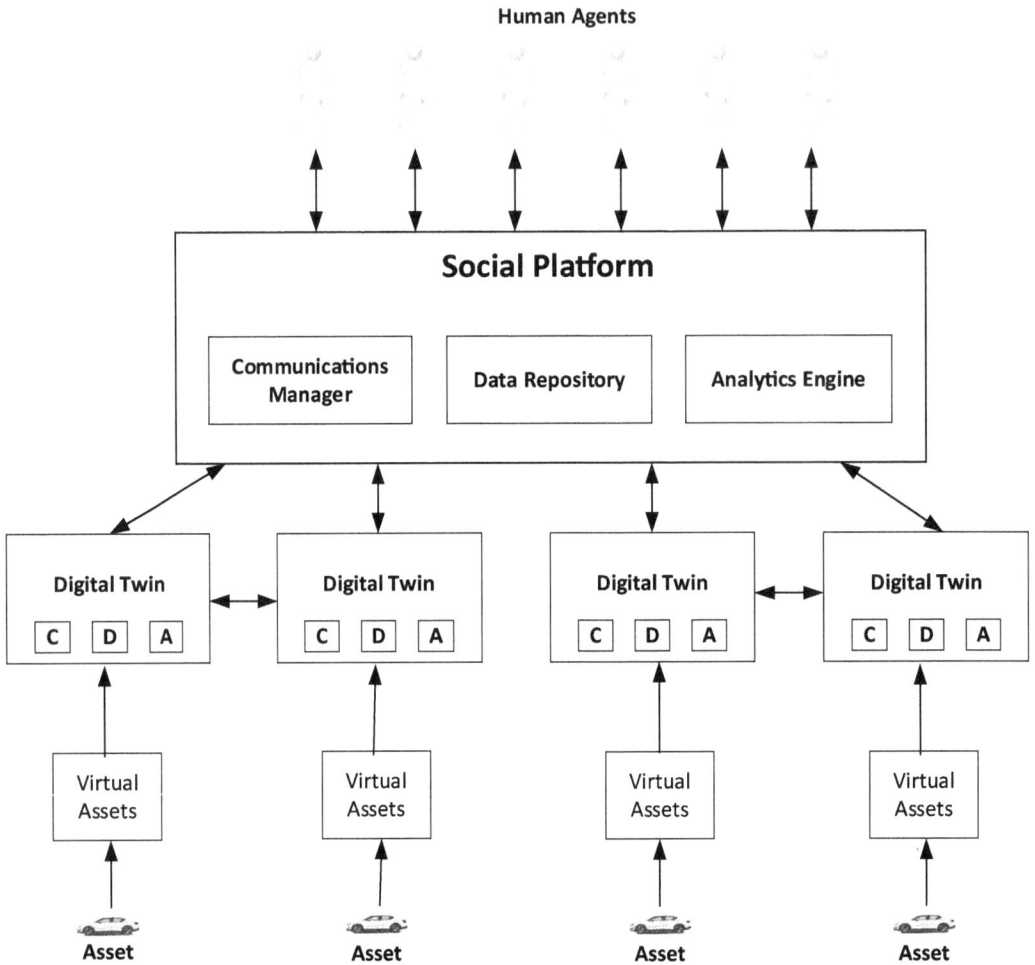

FIGURE 5.3 Heterarchical MAS Architecture.

prognostics, and the digital twins within the same cluster collaborate with one another. There is, however, an important difference: the clustering algorithm implemented here must be a distributed clustering algorithm, unlike the previous architectures, where the social platform performs this task. The distributed k-mean clustering algorithm implemented here is like the one presented in Qin et al. [1] and is detailed in "Distributed clustering algorithm." In the centralized and hierarchical case, the failure of a mediator agent or the social platform leads to a halt of predictive maintenance operations for hundreds of assets in the system, which then causes a dramatic increase of corrective maintenance actions. In the distributed and heterarchical cases, maintenance recommendations are produced by the digital twins, and no agent failure has the potential to compromise maintenance for hundreds of assets. The summary of analysis is described as follows:

- Higher costs for low-value assets across the board: the normalized cost per asset and time-step in architectures containing low-value assets is orders of magnitude higher than for the case of high-value assets, the cost for all architectures increases. This means that real-time MAS implementations for prognostics are more cost-effective, and the more expensive the replacement cost of the assets is if prognostics complexity remains constant.

Human Agents

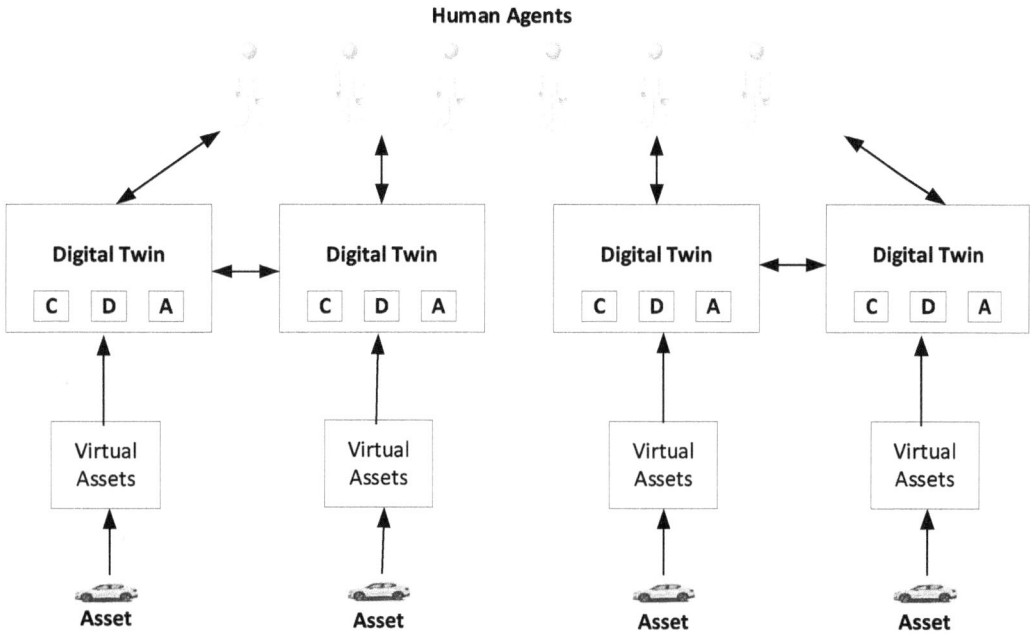

FIGURE 5.4   Distributed MAS Architecture.

- When there are no agent failures, centralized and hierarchical architectures are generally cheaper; this is expected, as these two architectures are also the ones featuring less communication and computation costs. Only for very noisy experiments, for very low communication costs or very high asset values this becomes false, when maintenance costs dominate over the rest of the costs of the system.
- When there are no agent failures, cost differences between architectures minimize as asset value increases. This is because in high-value assets maintenance costs dominate over communication and computing costs. If no agent failures are included in the experiments, predictive maintenance in the different architectures has a very similar level of accuracy, and the overall cost is essentially the same.
- A high communication cost limit exists: if communication costs are high enough (or asset value low enough), agent failures mean lower operational costs. The explanation for this is simple: agent failure increases operational cost through more unwanted corrective maintenance actions but decreases it by halting computation and communication actions. If the communication costs are high enough, agent failure then leads to a less costly architecture.

5.9   SUMMARY

We have studied the cost consequences of implementing different multi-agent system architectures for collaborative prognostics, a new prognostics approach based on collaboration between agents that represent different assets in the fleet. Four architectures are analyzed, featuring different levels of distribution: centralized, hierarchical, heterarchical, and distributed. The main conclusion drawn from this analysis is that decentralized architectures are not always cost-efficient for the purpose of collaborative prognostics. If the assets in the system have a low value, communication and computing costs become relevant, and more centralized architectures become the best option. However, when the value of the assets is high enough, the implementation of distributed architectures can be justified. In this case, the value of the assets is much larger than communication and computing costs, and the benefits of distributed architectures can be leveraged. This difference between architectures becomes especially relevant when agent failure is included. In this case, architectures where prognostics and maintenance planning are highly dependent on few agents are especially susceptible to agent failure. This, in practice, means that when agent failures are considered, distributed architectures become more competitive.

A secondary conclusion is that multi-agent-based collaborative prognostics architectures are more cost-efficient in general than the assets of the system are. This is a common-sense result: there is no point in enhancing very low value assets with IoT technologies, as the cost of these technologies outweighs by far the savings of a predictive maintenance policy. With regard to future work, there are some parametric dependencies that have not been explicitly studied in this chapter. Perhaps, the most important is the number of clusters k (groups of different assets) in the fleet, which in the experiments has been limited to four. In the distributed and hierarchical architectures, communication costs C_c will be proportional to the square of the size of each cluster: $C_c \propto \left(\dfrac{N}{k} \right)^2$ (if each cluster has a similar size). This means that if k is kept constant but total number of assets N is increased, the cost of distributed and hierarchical architectures will increase at a higher rate than the cost of their centralized and heterarchical counterparts. Studying the optimality of different architectures with respect to the heterogeneity of the fleet (given by k) would thus give place to a potentially interesting research study.

Another parametric dependency that has been omitted (kept constant) in the experiments is the dependence of the cost of the architectures with the probability of agent failure and its duration. In our experiments, the first parameter is kept constant, and the second is sampled from a predetermined probability distribution. We purposely chose both parameters to be relatively high to compensate for the fact that we assumed that there were no costs associated to repairing agent failure. Further research should focus on exploring this dependency and placing it within realistic industrial parameters. Finally, in the experiments presented here, the maintenance threshold η is kept constant. Although this is a reasonable assumption for the purpose of this particular example, comparing across architectures, studying the effect of optimizing in real time would yield a potentially interesting research study.

REFERENCES

[1] Palau, A. S. et al., "Multi-Agent System Architectures for Collaborative Prognostics," Journal of Intelligent Manufacturing. 30: 2999–3013, 2019, https://doi.org/10.1007/s10845-019-01478-9.
[2] Poole, D. L. and Mackworth, A. K., "Artificial Intelligence: Foundation of Computational Agents," Second Edition, 2017, Cambridge University Press.

6 Protocol Layer Architecture

6.1 OPEN STANDARD INTERNATIONAL MODEL

Most importantly, we can imagine an environment where thousands of AI supercomputers will be connected to billions of AI client devices over a global network. In this situation, AI SW/HW technologies will be viewed as another set of new communications networking technologies. It implies that we need to use the Open Standard Interconnection (OSI) model that provides the seven-layered protocol architecture for AI software and hardware technologies. The OSI model is developed by the International Standard Organization (ISO). The ISO/OSI Protocol Layer Model of Architecture [1] has ushered a revolutionary mechanism for providing interoperability among all functional entities connected across the network in multi-vendor products and services based on standards. Rightly, OSI Model has provided options for creating value-added new innovative products and services in the application (Layer 7) for fostering competitions.

Figure 6.1 shows the OSI Protocol Layer Architecture for end-to-end communications over the network. Seven logical protocol layer stacks are shown for the source/destination entity where computer host and user terminal/device are located. Interestingly, the network router/switch that connects the host and terminal/device over the network contains only up to three logical protocol stacks.

Basic principles of each layer have been that each layer is created to have different layer of abstraction distinguishing each other based on international standards. Each layer needs to perform a well-defined, standard-based, specific set of functions separating distinctly from each layer logically. The software interface termed as the application programming interface (API), not shown in Figure 6.1, can be defined between two layers for transparent communications between the two layers. The OSI itself is not a network architecture because OSI model does not specify the exact services and protocols for each layer. The exact services and protocols need to be standardized in the international standard body for providing interoperability. The details are not addressed here. Interested readers might see the OSI standard [1] of ISO. However, we provide a summary of some example functions of each OSI layer along with some examples of some specific functions that have been standardized later in different international standard forums.

6.1.1 LAYER 1: PHYSICAL LAYER

The physical layer is responsible for transmission and reception of raw bits over a communication link or channel. The physical layer link/channel can be physical wire or wireless. The traffic multiplexing, modulation scheme, data rate, channel access method, physical connectors, simplex, full/half-duplex, timing, frequency, voltage level, and others are the functions of physical layer. The Institute of Electrical and Electronics Engineers (IEEE) has become the primary standard forum to build specifications related to Physical Layer 1. Ethernet, Bluetooth, universal serial bus (USB) are some examples of physical layer standard entities. The physical layer link/channel can be physical wire or wireless.

6.1.2 LAYER 2: DATA LINK LAYER

The data link layer transfers and receives data frames between two nodes that are connected by a wire or wireless link. Error correction that occurs for transmission over the link and acknowledgment of data frames are primary functions. Like Layer 1, IEEE has also become the primary standard forum to build specifications related to Data Link Layer 2. IEEE has further extended the OSI

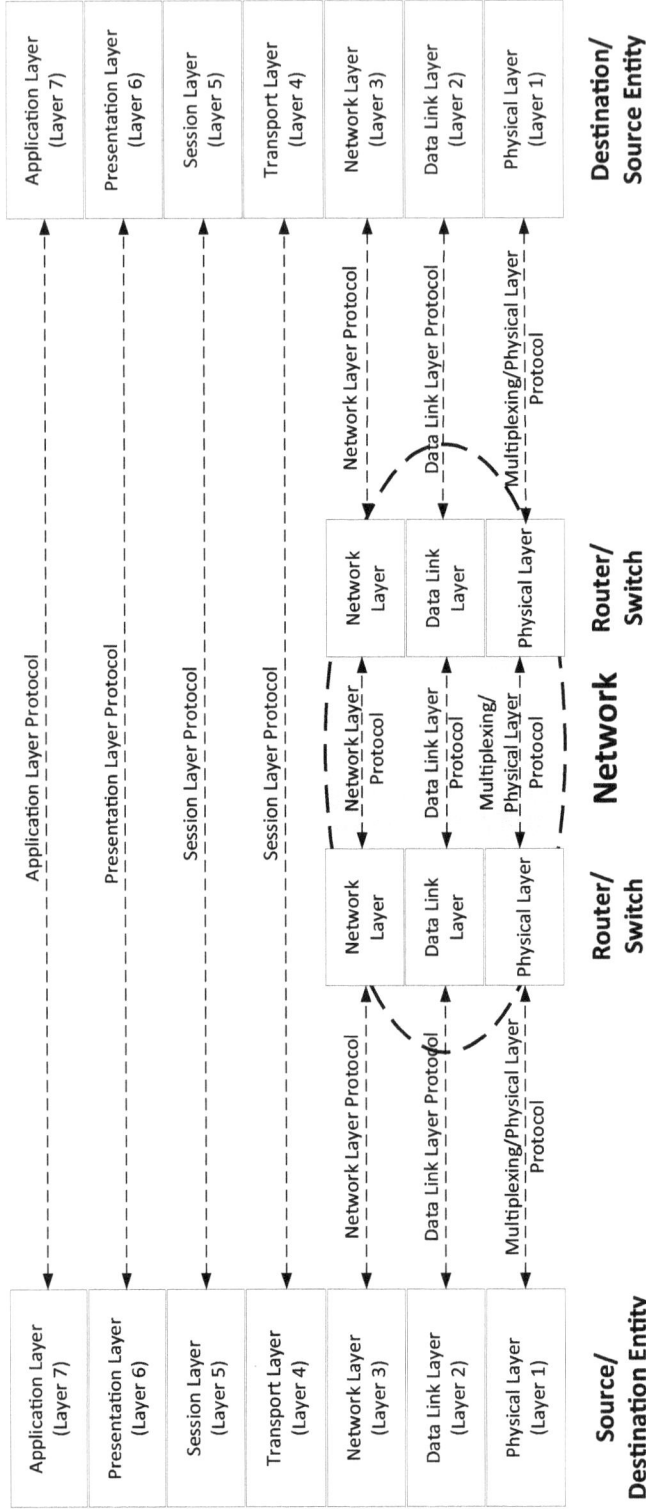

FIGURE 6.1 OSI Protocol Layer Reference Architecture.

Data Link Layer 2: Sublayer

Logical Link Control (LLC): Connection-Oriented/Connectionless
IEEE 802.2. Link Addressing, Sequencing, Service Access Points (SAPs)

(e.g. Asynchronous Transfer Mode (ATM), Frame Relay (FR), Point-to-Point
Protocol (PPP), Synchronous Optical Network (SONET)

Media Access Control (MAC)
Regulating Access to Transmission Media, Maintaining MAC Protocol Stack

(e.g. Ethernet - Carrier-Sense Multiple Access – Collision Detection (CSMA/CD)
Protocol, Wi-Fi – CSMA - Collision Avoidance (CSMA/CD) Protocol)

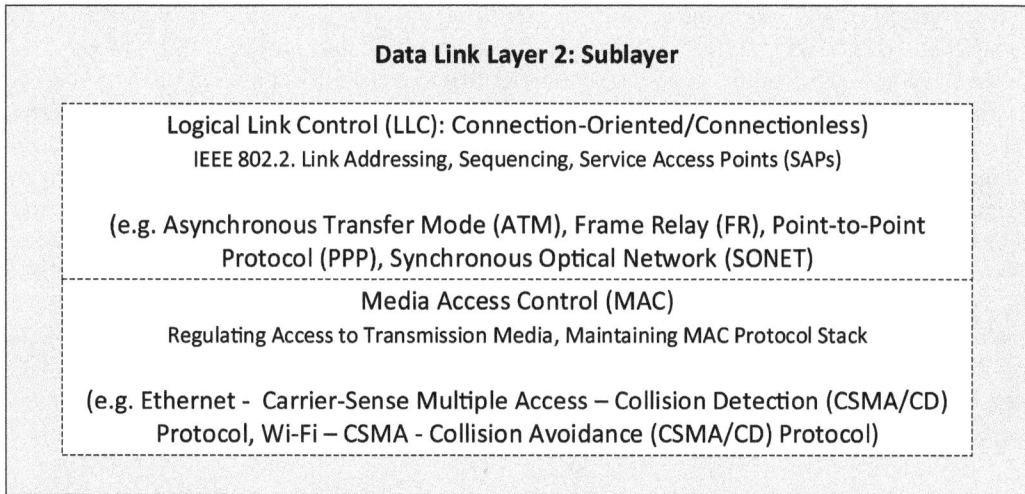

FIGURE 6.2 OSI Data Link Layer 2 Sublayer defined in IEEE: Logical Link Control (LLC) and Media Access Control (MAC).

Data Link Layer 2 by subdividing it into two sublayers: Media Access Control (MAC) and Logical Link Control (LLC) as shown in Figure 6.2. The LLC sublayer acts as an interface between the media access control (MAC) sublayer and the Network Layer 3. The LLC sublayer provides the logic for the data link and controls the synchronization, flow control, and error-checking functions of the data link layer. Again, LLC has the option to support both connection-oriented and connectionless protocol. With connection-oriented communication, each LLC frame that is sent is acknowledged while in the case of connectionless communications, no LLC frame needs acknowledgment.

IEEE has developed well-defined standards for LLC such as IEEE 802.2. In addition, Point-to-Point Protocol (PPP), Asynchronous Transfer Mode (ATM), Frame Relay (FR), and Synchronous Optical Network (SONET) standards developed in other standards forums also belong to LLC. In the same token, IEEE has developed MAC standards such as Ethernet: Carrier-Sense Multiple Access—Collision Detection (CSMA/CD) IEEE 802.3 for a variety of data rates, Wi-Fi: CSMA-Collision Avoidance (CSMA/CA) IEEE 802.11, and 802.15.4 Zig-Bee have also developed in the IEEE.

6.1.3 Layer 3: Network Layer

The network layer is concerned with controlling the operations of the subnetwork/network. A key function of the network layer protocol is how a packet is routed from the source to the destination that may be connected over the subnetwork/network many hops away from the source. A standardized network routing protocol (e.g., Open Short Path First (OSPF), Border Gateway Protocol (BGP), and others) needs to be used which next hop network node is in the best situated in the path to the destination node resolving the network layer address of the destination node. In addition, routing protocol, multicast group management, network-layer information and error, and network-layer address assignment are the functions of the network layer. Note that it is the function of the payload that makes these belong to the network layer, not the function of the protocol that carries them.

6.1.4 Layer 4: Transport Layer

The basic function of the transport layer situated between the session (Layer 5) and the network layer (Layer 3) is to accept data from the session layer, then split into smaller packets as permitted per the network layer, and pass to the network layer ensuring that the packets will arrive correctly (for connection oriented transport protocol such as Transmission Control Protocol (TCP) standardized in the IETF and others) at the other hand. It implies that the reliability of a given link is controlled by the transport layer through flow control, segmentation and de-segmentation, and error control. The connectionless transport protocol is used (e.g., User Datagram Protocol (UDP) standardized in the IETF and others), the reliability services of the packets are pushed to the upper application layer using some proprietary mechanisms, and flow control is also not offered by the transport layer.

Another transport protocol standardized in the IETF is the newer Stream Control Transmission Protocol (SCTP). STCP is also a reliable, connection-oriented transport mechanism like TCP. But SCTP is message-stream-oriented, not byte-stream-oriented like TCP, and provides multiple streams multiplexed over a single connection. The newer feature like multi-homing is also supported by the SCTP, in which a connection end can be represented by multiple IP addresses (representing multiple physical interfaces), such that if one fails, the connection is not interrupted. Other standard bodies have also standardized some transport protocols but have not been widely used in the industry.

6.1.5   LAYER 5: SESSION LAYER

The session layer allows users on different computers/machines to establish sessions between them controlling the dialog. It establishes, manages, and terminates the connections between the local and remote application. It provides functions such as full/half-duplex, or simplex operation, and establishes procedures for check-pointing, suspending, restarting, and terminating a session. Note that the Internet Protocol Layer stack (see Section in 6.2.2) that does not contain the session layer, the suspending and closing function of the session of connection-oriented transport protocol (e.g., TCP) are merged with the lower Transport Layer (Layer 4).

6.1.6   LAYER 6: PRESENTATION LAYER

The application layer entities between the two communicating entities may have different syntax and semantics. The presentation layer mitigates the differences of syntax and semantics through mapping through translation including compression and decompression. That is, the presentation layer transforms data into the form that the application layer accepts. This layer formats data to be sent across a network. For example, audio coding, video coding, mapping between non-XML to XML format, and others.

6.1.7   LAYER 7: APPLICATION LAYER

The application layer is where the end-user interacts directly with the software applications. OSI standard does not define this layer leaving to the market to develop intelligent and value-added applications and services through competitive means based on market demand fueled by customers throughout the world. Of course, there are opportunities to standardize some application layer protocols in the international standard forums such as SMTP for emails, Session Initiation Protocol (SIP) for Voice-over-Internet Protocol (VoIP), File Transfer Protocol (FTP) for file transfers, and others. It implies that the value-added services to be built on the top of the standardized application layer protocol is so huge that some sort of interoperable products in multi-vendor environments are needed for reducing costs building scalable global communications networks.

More precisely, there could be many building blocks in the application layer differencing between the application-entity and application itself. For example, application layer functions typically include identifying communication parties, determining resource availability, and synchronizing communication. When identifying communication partners, the application layer determines the identity and availability of communication parties for an application with data to transmit. For VoIP/multimedia communications, SIP enables exactly to solve these problems and establishes

VoIP/multimedia calls between the parties. It is interesting to note that the market value of value-added services built on the top of interoperable SIP standard-based products is also substantial [2–4]. The application layer is the most important area for building products and services using AI-enabled value-added applications and services.

6.2 INTERNET MODEL

The International Engineering Task Force (IETF) has developed an early Transmission Control Protocol (TCP)/Internet Protocol (IP) Reference Model that has evolved over the years. RFC 1122 defines the enhanced TCP/IP Reference Model titled as Host Requirements. The document loosely defined layer model without specifying any layer number as it has been done in the OSI model. Finally, IETF has not done any formal enhancement of their layer model while OSI model seems to be more logical. The layer model that has been popularized in the IETF evolving from RFC 1122 is depicted in Figure 6.3.

We will be using Figure 6.3 as the basis for describing the protocol layer reference architecture throughout the book because the Internet has become the most dominant global network for all users throughout the world. In addition, the detail of the Internet layer is shown in Figure 6.4 will be used in this book. In addition, numerous applications for both commercial and military arena are also Internet Protocol (IP)-based. IETF emphasizes for a principle to put intelligence at the edges and postulates that the Internet itself (other than edges) should be simple just to provide faster communications to the edge nodes and should not keep any state for simplicity. In reality, due to the concern that network Internet Protocol version 4 (IPv4) with 64-bit address field will not be able to support all the network addresses connecting all the devices that users will be connecting to the IPv4 Internet as well as to meet the customers' need for topology hiding of their private network, address translators (NATs) have been used. Later on, firewalls have been used for preventing cyberattacks. Finally, the advent of popular World Wide Web services using webservers have forced us to use content caches, and the growing popularity of cloud services have forced to modify the principle of the IETF to some extent. With the standardization of IP version 6 (IPv6) with 128-bit address field has eliminated the needs for using NATs, which has much more address space as well as topology-hiding capability.

Keeping this principle as basis by the IETF, the different layers have been defined although not as clearly as in the OSI model, it is found that the use of layer numbers like those of the OSI model is convenient and easy to express or when the software and product are developed. It appears that there is almost one-to-one mapping between the IETF and OSI layer model despite some differences. For example, IETF Network Layer 3 for IP is not exactly as defined by OSI Network Layer 3 because IP has some features like ICMP and others that seem to be above Layer 3. On the other hand, IETF model does not define explicitly Session Layer 5 and Presentation Layer 6.

Initially, IETF has assumed both layers 5 and 6 are a part of the application layer. Later on, a consensus has emerged in the industry that layers 5 and 6 can be combined terming as "Middleware Layer" as shown in Figure 6.3. This is because the middleware functional entities can be used by one another dynamically by applications without formally maintaining the hierarchical structure as defined in the OSI model (Figure 6.1). We will use the term middleware throughout the book although not formally defined by the IETF protocol layer model (Figure 6.3). In the same token, IETF has not defined the Physical Layer 1 although this layer is recognized in the industry as standard, and products have been developed strictly following the definitions of OSI model. This book will also use the same convention as shown in Figure 6.3.

6.2.1 LAYER 1: PHYSICAL LAYER

The Internet model does not define anything specifically for the physical layer as shown in Figure 6.3. So the physical layer defined by the OSI model shown in Figure 6.1 has been used as standardized

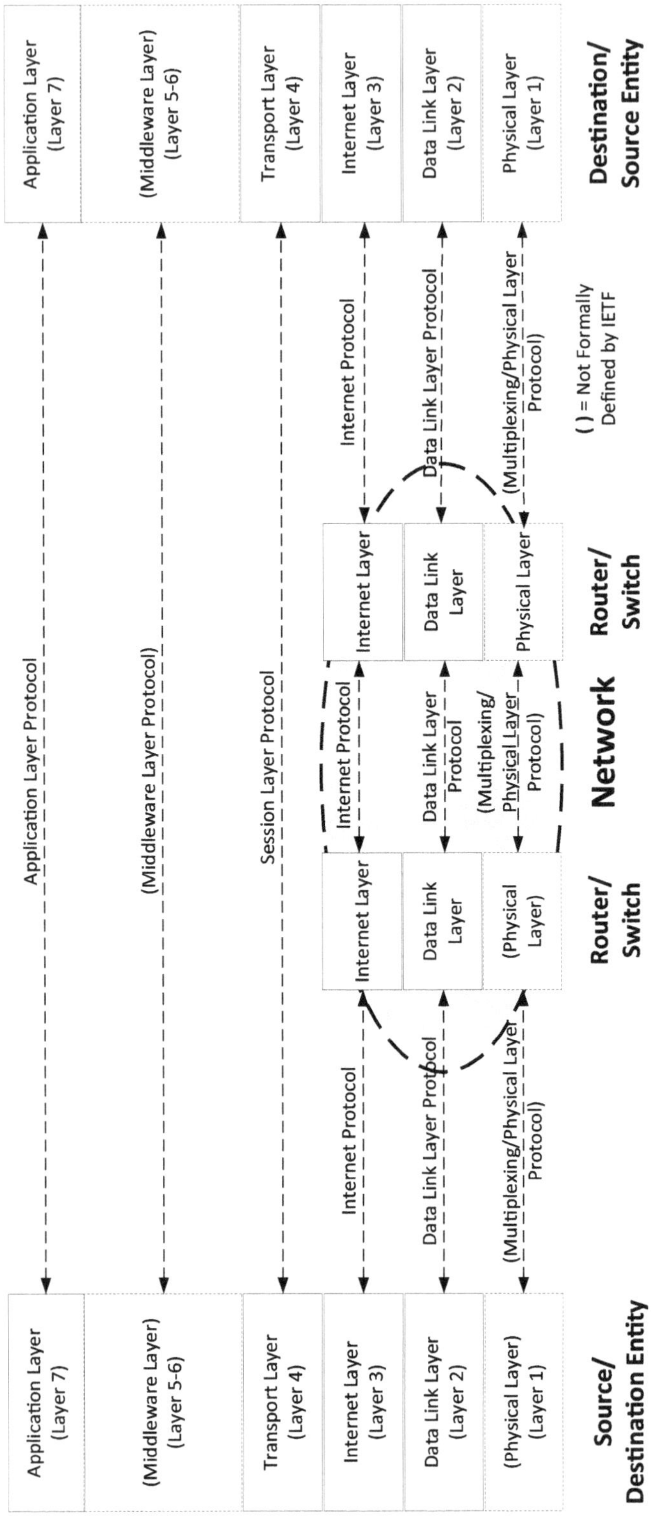

FIGURE 6.3 Internet Protocol Layer Reference Architecture.

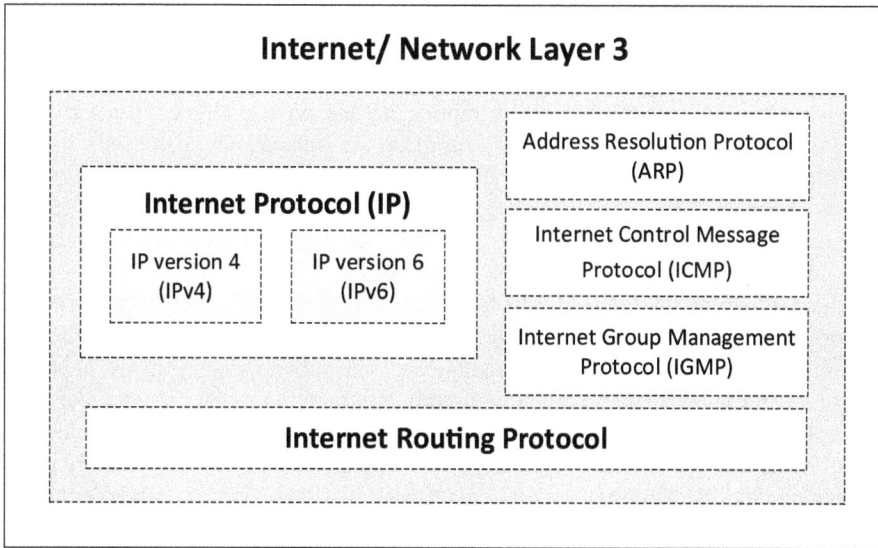

FIGURE 6.4 High-Level Detail of Internet/Network Layer 3.

specifications to define and build the products in the industry. IEEE has become the primary standard forum to build specifications related to Physical Layer 1.

6.2.2 LAYER 2: DATA LINK LAYER

The link layer of the Internet model corresponds to the Data Link Layer 2 of the OSI model. That is why we have named the link layer as the Data Link Layer (Figure 6.2) instead of the link layer. Moreover, the specifications of the OSI model (Figures 6.1 and 6.2) are also applicable for the Internet model (Figure 6.2). Again, IEEE is the main standard body that defines the specifications for this Layer 2. The description provided in the OSI model Data Link Layer 2 (Figures 6.1 and 6.2) are also applicable for the Internet model.

6.2.3 LAYER 3: INTERNET/NETWORK LAYER

The Internet layer is the key protocol layer that has been modeled specific to the connectionless Internet Protocol (IP) by the IETF. This layer is almost equivalent to OSI Layer Network Layer 3 (Figure 6.1) except that some simple address resolution and management-related protocols (Figure 6.4) such as Address Resolution Protocol (ARP), Internet Control Management Protocol (ICMP), and Internet Group Management Protocol (IGMP) seem to go beyond Layer 3 that OSI model envisioned. However, we will consider the Internet and Network Layer 3 as synonymous throughout the book as this term is popularly used in the industry. In the same token, the Internet model uses the term "gateway" for router, switch, or node, but industry has popularly used the term router, switch, or node and not the gateway. We would do the same in this book. The term Internet itself indicates that the IP protocol is used to interconnect the subnets that implies interworking between the subnetworks and exchange datagram messages between the network boundaries with uniform interface using hierarchical IP network addresses hiding the network topology. The internet that we know today connects every single node or device that is identified using the IP address. The IP frame format includes data fields for checksums and IP network addresses (64-bit for IPv4 and 128-bit for IPv6).

IETF has developed many routing protocols over the years for routing the IP packets over the Internet. For example, OSPF is a popular routing protocol using a link state routing (LSR) and has primarily been used in the subnetwork that falls into the group of "interior" gateway protocol in IETF terminology, operating within a single autonomous system (AS). BGP is another important routing protocol that has been standardized as the "exterior" gateway protocol designed to route messages among the autonomous systems. It implies that the network that connects all the autonomous systems will be using the BGP routing protocol. An analogy that can used to simplify the usage of these two routing protocols is that OPSF is used in the access network while the BGP is used in the backbone network. Another difference is that OSPF can be used dynamically to route packets through different paths based on the network traffic conditions while BGP, unlike OSPF, needs to be configured statically to create the routing paths.

The additional features that the Internet layer has include the ARP, ICMP, and IGMP. ARP is used by nodes to resolve corresponding physical addresses (e.g., MAC addresses) to their allocated IP addresses (e.g., IPv4 addresses). ICMP provides for some network management and maintenance functions, among these is the "ping" function used by nodes to query the properties of other nodes in the network. IGMP manages the membership of hosts and routing devices in multicast groups. IP hosts use IGMP to report their multicast group memberships to any immediately neighboring multicast routing devices. Multicast routing devices use IGMP to learn, for each of their attached physical networks, which groups have members.

6.2.4   LAYER 4: TRANSPORT LAYER

Transport layer 4 is designed, as described in the case of OSI model, that the transport protocols provide host-to-host communication services for applications. The description provided in the OSI Layer 4 is also applicable for the Internet model.

6.2.5   LAYER 5–6: MIDDLEWARE

The Internet model has not defined the middleware layer (Layer 5–6) other than making this as a part of the 5-Layer model. However, the industry consensus has been built over the years to have a middleware layer that consists of both layers 5 and 6 per OSI model (Figure 6.1). In fact, the middleware layer is the key layer for which a huge number of software components have been standardized in different industry forums and in each individual company worldwide. Middleware layer of the evolving Internet Model (encompassing both Layers 5 and 6 of OSI model) consists of somewhat standardized software modules developed in some manner in the industry, which is independent of the underlying low-layer hardware and software components, and makes it easier for the software developers to implement communication and input/output so that they can focus on the specific purpose of their application successfully achieving desired results within the cost, schedule, and complexity goals. In this respect, middleware is an abstraction layer that acts as an intermediary. Middleware manages interactions between application software and the underlying system software layers, such as the operating system and device driver layers. In addition, one middleware component can utilize services of another piece of middleware component and vice versa. The underlying software can be the embedded software that controls specific functions of the non-computer devices like microchip or as part of another application that sits on top of the chip.

Middleware software components can offer many benefits to the overlying middleware and applications: adaptive to changing availability of system resources, flexibility and scalability configuring and customizing the software and hardware limitations, security enhancement ensuring authorized access to system resources, portability avoiding time-consuming and expensive rewrites of the application code, and connectivity and inter-communications ability to transparently communicate with other applications on a remote device through some user-friendly, standardized interface. Examples of industry standardized middleware software components are Message Oriented Middleware (MOM), Object Request Broker (ORB), Enterprise Service Bus (ESB), Open Database Connectivity (ODBC), Java Database Connectivity (JBDC), Transaction Processing Monitor (TPM),

Virtual Machine (VM), web middleware, server middleware, and others. Although, by definition, middleware sits between the operating system (OS)/underlying (embedded) software and the application layer, both OS and underlying (embedded) software can also be considered as a kind of the middleware in principle.

In the beginning of the year 2000, the middleware development has taken an explosive turn due to rise of web services over the Internet along with cheap cloud computing throughout the world with the emergence of Software Oriented Architecture (SOA) framework defined by the Open Group (see Section 1.3). In the context of AI, we will introduce many other new intelligence middleware software components in this book. We will see that AI services that are used across for creation of powerful intelligent services using artificial agents can be through as kind of common of set middleware services for the value-added intelligent services.

6.2.6 LAYER 7: APPLICATION LAYER

The Application Layer (Layer 7) of the Internet model is basically the same what we have described in the case of the OSI model. A new generation of AI applications will be introduced in this book.

6.3 PEER-TO-PEER ARCHITECTURE MODEL

Earlier architectural models that we have described are known as the client-server (C/S) models where each client is a slave to the servers. In peer-to-peer (P2P) architecture, each node (e.g., computer) acts as an equal peer where each peer has equal capabilities. Peer-to-peer architecture (P2P architecture) is a commonly used computer networking architecture in which each workstation, or node, has the same capabilities and responsibilities. In contrast, P2P is compared to the classic client/server architecture in which some computers/nodes are dedicated to serving others. P2P may also be used to refer to a single software program designed so that each instance of the program may act as both client and server, with the same responsibilities and status. P2P architecture is often referred to as a peer-to-peer network. The content distribution is the key purpose of the P2P network. This also includes software publication and distribution, content delivery networks, streaming media, and peer casting for multicasting streams, which facilitate on-demand content delivery. Scientific research networks, search and communications networks, and modern dynamic unmanned aerial vehicles (UAVs)/drone networks are also a part of the P2P network.

P2P applications are one of the core issues in the controversy over network neutrality. It implies no restrictions on Internet content, format, technologies, equipment, or modes of communication where regulations (e.g., Internet content provider, governments) are difficult to apply. On the other hand, service providers or governments are attempting to control Internet use and content by directing network structure toward a client/server architecture. This sets up financial barriers for individuals and small publishers looking to gain Internet entry and sets up inefficiencies for sharing large files. There are three models of unstructured P2P computer network architecture: pure P2P, hybrid P2P, and centralized P2P.

In pure P2P system, there is a structured P2P computer network architecture, workstations (peers), and sometimes resources as well. They are organized according to specific criteria and algorithms. This leads to overlays with specific topologies and properties. In hybrid P2P setting, some weaknesses of the P2P system are augmented using client/server architecture. For example, P2P networks have clients with resources such as bandwidth, storage space, and processing power. This requires enormous amounts of resources in terms of security and file verifications making P2P system resistant to any kind of cyberattacks. By comparison, a typical client/server network shares demands but not resources. As a result, a hybrid system in combination of P2P and client/server will work better in this situation. In centralized P2P system, a centralized server takes care of each peer to determine the desired peers that contain the queried resource and then sends back the result to the requesting peer.

Figure 6.5 depicts the standard-based P2P architecture model taking P2P blockchain (BC) application as an example along with relationship non-BC P2P architecture model. P2P BC application has two distinct parts: P2P Transaction and Scripts and P2P BC Consensus & Mining Signaling Protocol. Non-BC P2P Applications have also two parts known as non-BC messages and non-BC signaling and messages. For example, some people are building ChatGPT P2P network. If ChatGPT uses standard-based P2P protocol as depicted in Figure 6.5, it can interoperate with other P2P products over the network. Note that the message transport and storage and logical topology plug-in (that uses application layer distributed hierarchical table) and forwarding and link management protocol remain the same for both BC and non-BC P2P applications.

A distributed hash table (DHT) is a decentralized hash table-based distributed topology plug-in system for mapping keys to values which provide lookup, storage schemes, and storing key-value pairs. Each node in a DHT is responsible for keys along with the mapped values. Any node can efficiently retrieve the value associated with a given key. DHT maps what the user is looking for to the peer that is storing the matching content. DHT could have a huge table that stores *who* has *what* data. Examples of DHTs are Chord, Kademlia, BitTorent, Dynamo, and others. Scalability is the main reason for choosing different DHTs for P2P networks.

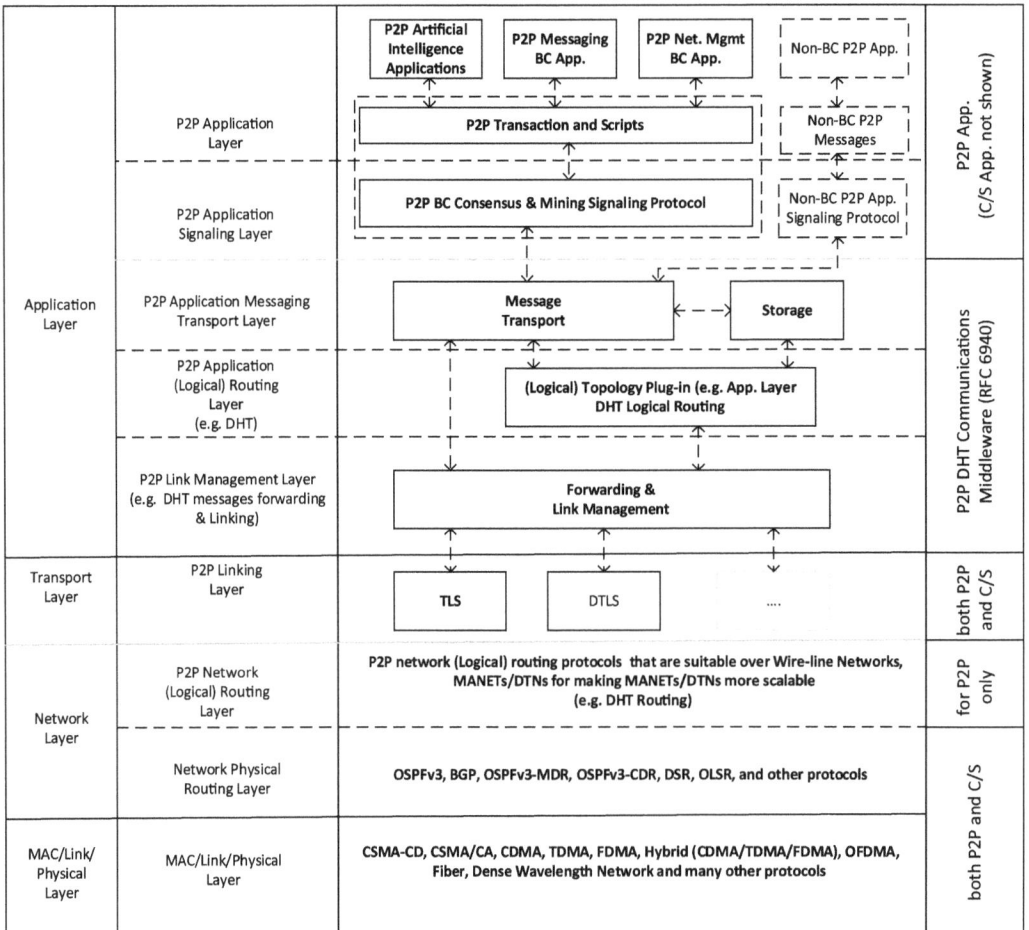

FIGURE 6.5 P2P Protocol Architecture Model.

However, the transport layer, network layer, and link/MAC layer remain the same for both P2P and non-P2P applications. Recently, people are also building ChatGPT P2P Applications networks. Note that ChatGPT is still considered as a set of the non-BC P2P applications that are using large natural language processing (NLP) protocol.

A mobile ad hoc network (MANET) is also a P2P network and needs two layers of routing protocols: Logical DHT-based Routing Protocol and Physical Routing Protocol. In this way, a large-scale MANET network can be built. The delay tolerant networks (DTNs) are also MANETs, but the DTN protocol is designed in such a way the delay of the messages over the network can tolerate disconnections of the links/nodes due to random movement of nodes from time to time. DSR, OLSR, and OSPFv3-MDR/CDR are some examples of MANET physical routing protocols on the top of which logical DHT routing protocols are used.

Border Gateway Protocol (BGP) is used for routing within the backbone network of the Internet while all other routing protocols described earlier are used in the access networks. CSMA/CD, CSMA/CA, CDMA, FDMA, Hybrid (CDMA/TDMA/FDMA), OFDMA, and others are used in the link/MAC Layer, which are known as the multiple access schemes. Fiber and Time/Dense Wavelength Network are used in the physical layer of the network, and we have described it in detail earlier as well as in the later sections.

The key point is that AI/ML/DL can be used in all these technologies in each layer of the protocol architecture to make them more efficient and improve performances.

6.4 SUMMARY

In this section, we have described the client-server (C/S) OSI protocol layer architecture, C/S Internet protocol layer model, and peer-to-peer (P2P) architecture model. The salient differences among these architecture models are also discussed in detail. We have indicated that Layers 4 and 5 of OSI model has been termed as "middleware" in the Internet model. The C/S model is a subset of the P2P protocol architecture model.

REFERENCE

[1] Zimmermann, H., "OSI Reference Model—The ISO Model of Architecture for Open Systems Interconnection," IEEE Transactions on Communications. 28 (4): 425–432, April 1980.

7 Artificial Intelligence Performance Analysis

At a very high level, artificial intelligence (AI) is a kind of agent-based allocation(s) that uses machine learning (ML) and/or deep learning (DL) algorithms to automate (i.e., self-service of the agent) services of application(s) with enhanced productivity/efficiency and reliability. Figure 7.1 depicts a high-level view of relationship between AI, ML, and DL.

Data is inputted to algorithms, which are aggregated in a certain relevant context (e.g., classification like modulations, channel access protocols, routing protocols, transport protocols, messaging protocols, audio/video coding, text, audio conferencing, videoconferencing, cybersecurity, command & control, common operating picture (COP), and management & operations) for AI application (say, cyberattack classification application—kinds of cyberattacks), in turn, leads to quicker and more accurate predictions by the AI agent (i.e., the AI agent of the cyberattack classification application—kinds of cyberattacks).

Machine learning uses algorithms to parse data into certain contexts (e.g., kinds of cyberattacks), learn from that data, and make informed decisions based on what it has been learned. Deep learning is a subfield of machine learning. DL structures algorithms in layers to create an "artificial neural network (ANN)" that can learn and make intelligent decisions (e.g., predictions of kinds of cyberattacks) on its own. Both ML and DL fall under the broad category of AI, but deep learning is what powers the most human-like AI.

However, the overall process of the AI/ML/DL remains uniform under all circumstances. Figure 7.2 shows an overview of the overall processing technique from signal/data acquisition to digitization to preprocessing, feature extraction/dimensionality reduction, feature classification through prediction, and finally output signal/data evaluation. Learning/training is done by the models before applying for prediction One of the important things of an artificial (AI) agent is acting behind to lead the learning process of AI/ML/DL techniques from feature extractions to dimensionality reduction to classification to prediction.

At first, the AI agent is trained to learn particular application (e.g., classification like modulations, channel access protocols, routing protocols, transport protocols, messaging protocols, audio/video coding, text, audio conferencing, videoconferencing, cybersecurity, command & control, common operating picture (COP), and management & operations) processes from feature extractions to dimensionality reduction to classification to prediction using a form of learning mechanism such as supervised learning using known labeled signal/datasets. Then unknown unlabeled signal/datasets are used for labeling with appropriate classification (e.g., kinds of cyberattacks).

The overall text classification process and the corresponding models that are used consists of the following steps:

- Signal/Data Acquisition
  - Non-ML/DL Algorithms (usual case) but could also use ML/DL in the future.
- Signal/Data Digitization
  - Non-ML/DL Algorithms (usual case) but could also use ML/DL in the future.
- Signal/Data Clearing and Preprocessing
  - Machine Learning
  - Deep Learning

 DOI: 10.1201/9781003499466-7

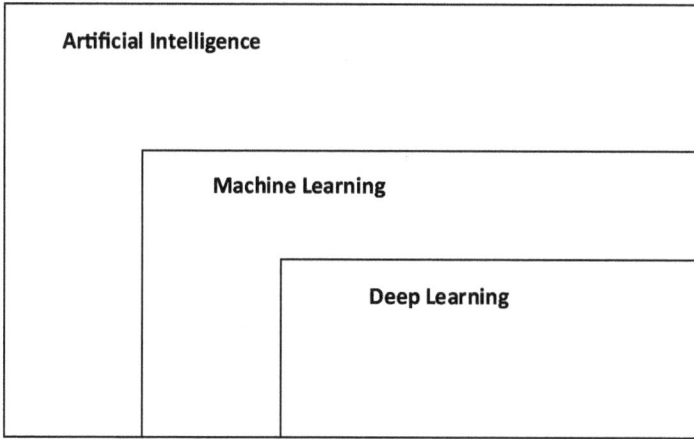

FIGURE 7.1 High-Level View of Relationship between Artificial Intelligence, Machine Learning, and Deep Learning.

- Feature Extraction
  - Machine Learning
  - Deep Learning
- Dimensionality Reduction
  - Machine Learning
  - Deep Learning
- Signal/Data Classification Techniques
  - Machine Learning
  - Deep Learning
- Signal/Data Classification Evaluation for Decision
  - Test Data Prediction
  - ML/DL Models Evaluation

7.1 SIGNAL/DATA ACQUISITION

The signal/data acquisition (SAQ/DAQ) is the process of acquiring information from a system in the form of digital form through sampling, for example. A sensor SAQ/DAQ system converts physical parameters to electrical signals. The SAQ/DAQ system, like signal conditioning circuitry, converts sensor signals into a form that can be converted to digital values. Analog-to-digital converters convert conditioned sensor signals to digital values. At this point, we are not making SAQ/DAQ system AI/ML/DL-enabled.

7.2 SIGNAL/DATA DIGITIZATION

Signal or data is sampled for digitization. For example, a sensor's physical parameters are converted into electrical signals that could be in analog form. We are assuming that all signal/data is digitized for use by the computer system. Once digitized information is available, the AL/ML/DL processes can be started.

7.3 SIGNAL/DATA CLEARING AND PREPROCESSING

The signal/data clearing and preprocessing is used primarily to remove noises and processes that are used for taking noises out will depend on the types of applications. For example, for text (e.g., emails, files, web pages), Tokenization, Stop Words, Capitalization, Slang and Abbreviation, Noise

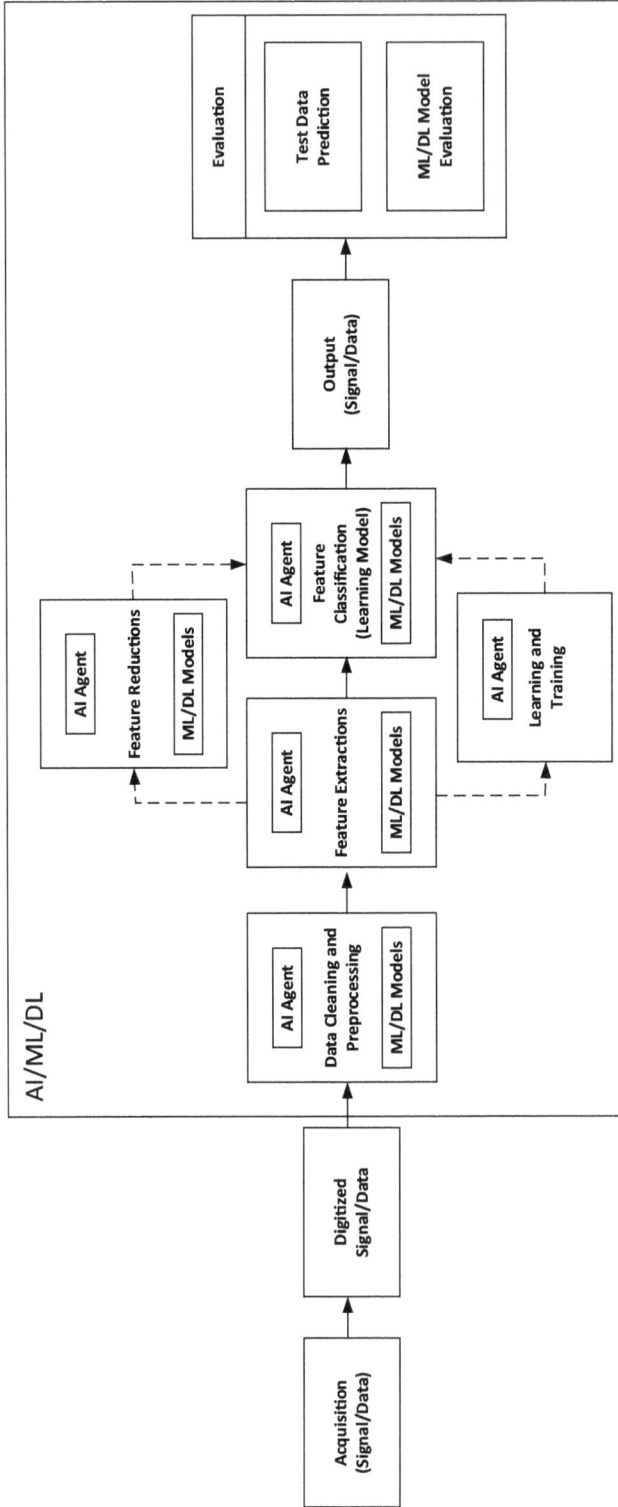

FIGURE 7.2 Overview of AI/ML/DL-based Preprocessing, Feature Extraction, and Feature Classification Process.

Removal, Spelling Correction, Stemming, and Lemmatization. Each of these functions has certain machine learning and non-ML algorithms.

7.4 FEATURE EXTRACTION

In machine/deep learning, the initial set of actionable non-redundant features (values) that have been extracted from the measured data used to meet the desired objectives such as facilitating the subsequent learning and generalization step, and in some cases, it might lead to better human understanding. The initial set of features is called the full set of features with no redundant features. Many ML feature extraction algorithms are available. However, what feature extraction algorithms need to be selected will again depend on the kind of applications. For example, text (e.g., emails, files, web pages) the following feature extraction algorithms can be used with each one with their pros and cons:

- Syntactic Word Representation
  - N-Gram
  - Syntactic N-Gram
- Weighted Words
  - Bag-of-Words (BoW)/Bag-of-Feature (BoF)
  - Limitation of Bag-of-Words (BoW)
  - Term Frequency-Inverse Document Frequency (TF-IDF)
- Word Embedding
  - Word-to-Vector (Word2Vec)
    - Continuous Bag-of-Words Model
    - Continuous Skip-Gram Model
  - Global Vectors for Word Representation (GloVe)
  - Fast Text (FastText)
  - Contextualized Word Representations

Note that any one of these algorithms will be suitable for text application feature extractions. One might use one or more algorithms for text application feature extraction. A usual case is to use as many algorithms as possible and select the best algorithm that yields superior results.

7.5 DIMENSIONALITY REDUCTION

Once the full set of non-redundant features is available, we have to identify the useful features and then select final feature set from the useful feature set for any applications including text. It is also a trial-and-error method to reduce the dimensionality of the features that are absolutely useful minimum feature set. Many general dimensionality reduction ML/DL algorithms are available as follows:

- Component Analysis
  - Principal Component Analysis (PCA)
  - Independent Component Analysis (ICA)
- Linear Discrimination Analysis (LDA)
- Non-Negative Matrix Factorization
- Random Projection
  - Random Kitchen Sinks
  - Johnson–Lindenstauss Lemma
- Autoencoder
  - Conventional Autoencoder Architecture
  - Recurrent Autoencoder Architecture
- t-distributed Stochastic Neighbor Embedding (t-SNE)

Dimensionality reduction is mostly used for improving computational time and reducing memory complexity:

- PCA attempts to find orthogonal projections of the dataset which contains the highest variance possible to extract linear correlations between variables of the dataset.
  - The main limitation of PCA is the computational complexity of this technique for dimensionality reduction.
    - To solve this problem, scientists introduced a random projection technique.
- LDA is a supervised technique for dimension reduction that can improve the predictive performance of the extracted features.
  - However, LDA requires researchers to manually input the number of components, requires labeled data, and produces features that are not easily interpretable.
- Random Projection is much faster computationally than PCA.
  - However, this method does not perform well for small datasets.
- Auto-Encoders require more data to train than other dimensionality reduction (DR) methods and thus cannot be used as a general-purpose dimensionality reduction algorithm without sufficient data.
- t-SNE is mostly used for data visualization in text and document datasets.

7.6 SIGNAL/DATA CLASSIFICATION TECHNIQUES

The most important step of the ML/DL process is choosing the best classifier. Without a complete conceptual understanding of each algorithm, one cannot effectively determine the most efficient model for classification of any applications. The most popular techniques for feature classification are listed here. First, we cover traditional methods of feature classification for any applications including text, such as the following:

Classification Techniques using Machine Learning (ML):

- Rocchio Classification
- Boosting and Bagging
  - Boosting
  - Bagging
- Logistic Regression (LR)
  - Combining Instance-Based Learning and Logistic Regression (LR)
  - Multinomial Logistic Regression
- Naïve Bayes Classifier
  - High-Level Description of Naïve Bayes Classifier
  - Multinomial Naïve Bayes Classifier
  - Naïve Bayes Classifier for Unbalanced Classes
- K-Nearest Neighbor (KNN)
  - Weight Adjusted K-Nearest Neighbor Classification
- Support Vector Machine (SVM)
  - Binary-Class SVM
  - Multi-Class SVM
  - String Kernel
  - Stacking Support Vector Machine (SVM)
  - Multiple Instant Learning (MIL)
- Design Tree
- Random Forest
  - Voting
- Conditional Random Field (CRF)

Feature Classification Techniques for any applications including text Using **Deep Learning (DL):**

- Deep Neural Networks
- Recurrent Neural Network (RNN)
  - Long Short-Term Memory (LSTM)
  - Gated Recurrent Unit (GRU)
- Conventional Neural Networks (CNN)
- Deep Belief Network (DBN)
- Hierarchical Attention Networks (HAN)
- Combination Techniques
  - Random Multi-Model Deep Learning (RMDL)
  - Stochastic Gradient Descent (SGD) Optimizer
  - Root Square Mean Proportionate (RMSProp) Step Size Optimizer
  - Adam Optimizer
  - Adaptive Gradient Algorithm (*Adagrad*) Optimizer
  - Adaptive Learning Rate per Dimension (Adadelta) Optimizer
  - Hierarchical Deep Learning for Text (HDLTex)
  - Other Techniques
    - Recurrent Convolutional Neural Networks (RCNN)
    - CNN with LSTM (C-LSTM)

The most important step of the text classification pipeline is choosing the best feature classifier. Without a complete conceptual understanding of each algorithm, one cannot effectively determine the most efficient model for a feature classification application. Non-parametric techniques have been studied and used as classification tasks such as k-nearest neighbor (KNN). Support Vector Machine (SVM) is another popular technique which employs a discriminative classifier for document categorization. This technique can also be used in all domains of data mining, such as bioinformatics, image, video, human activity classification, safety, and security, and so on. This model is also used as a baseline for many researchers to compare against their own works to highlight novelty and contributions.

Tree-based classifiers such as decision tree and random forest have also been studied with respect to document categorization. Each tree-based algorithm will be covered in a separate subsection. In recent years, graphical classifications have been considered as a classification task such as conditional random fields (CRFs). However, these techniques are mostly used for document summarization and automatic keyword extraction.

Lately, deep learning approaches have achieved surpassing results in comparison to previous machine learning algorithms on tasks such as image classification, natural language processing, face recognition, and so on. The success of these deep learning algorithms relies on their capacity to model complex and nonlinear relationships within data. The key is that one has to select a minimum set of algorithms to test their applications for better classification before taking the final decision.

7.7   SIGNAL/DATA CLASSIFICATION EVALUATION CRITERIA

The final part of the classification pipeline is evaluation for any applications including text. Understanding how a model performs is essential to the use and development of text classification methods. There are many methods available for evaluating supervised techniques. Accuracy calculation is the simplest method of evaluation but does not work for unbalanced datasets. The following evaluation methods for text classification algorithms are used:

- Confusion Matrix
- Macro- and Micro-Averaging

- F_b-Score
- Matthews Correlation Coefficient (MCC)
- Receiver Operating Characteristics (ROC)
- Area under the ROC Curve (AUC)

Unlike classical method, the ML/DL technique requires the results should be tested by a set of two or more performance evaluation algorithms through comparisons before selecting the final best prediction.

7.7.1 CONFUSION MATRIX

The confusion matrix is a tool for predictive analysis in machine learning. To check the performance of a classification-based machine learning model, the confusion matrix is deployed. We can also say Confusion matrix is a summarized table of the number of correct and incorrect predictions yielded by a classifier (or a classification model) for binary classification tasks. A Confusion matrix is an $N \times N$ matrix (Figure 7.3) used for evaluating the performance of a classification model, where N is the number of target classes.

By visualizing the confusion matrix, an individual could determine the accuracy of the model by observing the diagonal values for measuring the number of accurate classifications. The confusion matrix is in the form of a square matrix where the column represents the actual values, and the row depicts the predicted value of the model and vice versa.

- **TP: True Positive:** The actual value was positive, and the model predicted a positive value.
- **FP: False Positive:** Your prediction is positive, and it is false. (Also known as the Type 1 error).
- **FN: False Negative:** Your prediction is negative, and the result is also false. (Also known as the Type 2 error).
- **TN: True Negative:** The actual value was negative, and the model predicted a negative value.

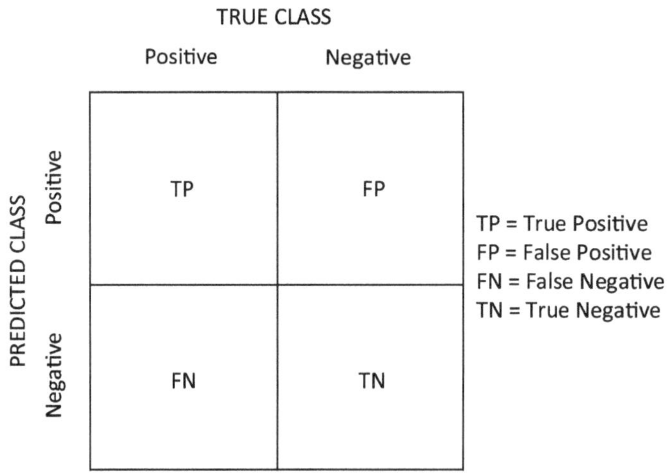

FIGURE 7.3   Confusion Matrix.

- **Accuracy:** Accuracy is a measure for how many correct predictions your model made for the complete test dataset. Accuracy is a good basic metric to measure the performance of the model. In unbalanced datasets, accuracy becomes a poor metric.

$$Accuracy = \frac{TP + TN}{TP + TN + FP + FN}$$

$$Misclassification = \frac{FP + FN}{TP + TN + FP + FN}$$

- **Precision:** Precision tells us how many of the correctly predicted cases actually turned out to be positive. This would determine whether our model is reliable or not.

$$Precision = \frac{TP}{TP + TN}$$

Precision is a useful metric in cases where False Positive is a higher concern than False Negative.
- **Recall (Sensitivity):** Recall tells us how many of the actual positive cases we were able to predict correctly with our model. Recall is a useful metric in cases when False Negative trumps False Positive.

$$Recall = \frac{TP}{TP + FN}$$

A higher recall means that most of the positive cases (TP+FN) will be labeled as positive (TP). This will lead to a higher number of FP measurement and a lower overall accuracy. A low recall means that one has a high number of FN (should have been positive but labeled as negative). This means that one has more certainty if you found a positive case, this is likely to be a true positive.
- **F1-Score:** When we try to increase the precision of model, the recall grows down and vice versa. F1-Score is a harmonic mean of Precision and Recall, and so it gives a combined idea about these two metrics. It is maximum when precision is equal to recall.

$$F1 - Score = 2 * \frac{1}{\dfrac{1}{Precision} + \dfrac{1}{Recall}} = 2 * \frac{Precision * Recall}{Precision + Recall}$$

The interpretability of the F1-Score is poor, it means that we don't know what our classifier is maximizing—precision or recall. So we use it in combination with other evaluation metrics which gives us a complete picture of the result.
- **False Positive Rate:** False positive rate (FPR) is a measure of how many results get predicted as positive out of all the negative cases.

$$FPR = \frac{FP}{TN + FP}$$

7.7.2 MACRO- AND MICRO-AVERAGING

The difference between macro- and micro-averaging is that macro-averaging gives equal weight to each category while micro- averaging gives equal weight to each sample. If we have the same number of samples for each class, both macro and micro will provide the same score [1].

7.7.2.1 Micro-Average Precision

The micro-average precision is the sum of all true positives divided by the sum of all true positives and false positives. In other words, we divide the number of correctly identified predictions by the total number of predictions. The micro-average calculation of the earlier data would be as follows:

$$PrecisionMicroAvg = \frac{\left(TP_1 + TP_2 + \ldots + TP_n\right)}{\left(\left(TP_1 + TP_2 + \ldots + TP_n\right) + \left(FP_1 + FP_2 + \ldots + FP_n\right)\right)}$$

7.7.2.2 Macro-Average Precision

The macro average precision is the arithmetic mean of all the precision values for the different classes. The macro-average calculation of the earlier data would be as follows:

$$PrecisionMacroAvg = \frac{\left(Precision_1 + Precision_2 + \ldots + Precision_n\right)}{n}$$

7.7.2.3 Micro-Average Recall

The micro-average recall is the sum of true positives for all classes divided by actual positives (rather than predicted positives). The micro-average calculation of the earlier data would be as follows:

$$RecallMicroAvg = \frac{\left(TP_1 + TP_2 + \ldots + TP_n\right)}{\left(\left(TP_1 + TP_2 + \ldots + TP_n\right) + \left(FN_1 + FN_2 + \ldots + FN_n\right)\right)}$$

7.7.2.4 Macro-Average Recall

The macro-average recall is the arithmetic mean of all recall scores for different classes. The macro-average calculation of the earlier data would be as follows:

$$RecallMacroAvg = \frac{\left(Recall_1 + Recall_2 + \ldots + Recall_n\right)}{n}$$

Here is what we can infer about micro- and macro-averages:

7.7.2.5 Micro-Averaging

Micro-average statistics evaluate models trained for multi-class classification problems. We use micro-averaging when we need an equal weight for each instance or prediction. The micro-average precision is the sum of true positives for a single class divided by the sum of predicted positives for all classes. The micro-average recall score is the sum of true positives for a single class divided by the sum of true positives for all classes.

7.7.2.6 Macro-Averaging

Macro-average statistics evaluate models trained for multi-class classification problems. We use macro-averaging in case of a class unbalance (different number of instances associated with different class labels). The macro-average is the arithmetic mean of the individual class related to precision, memory, and f1 score. We use macro-average scores when we need to treat all classes equally to evaluate the overall performance of the classifier against the most common class labels.

7.7.3 F_β-Score

In statistical analysis of binary classification, the F-score or F-measure is a measure of a test's accuracy. It is calculated from the precision and recall of the test, where the precision is the number of true positive results divided by the number of all positive results, including those not identified correctly, and the recall is the number of true positive results divided by the number of all samples that should have been identified as positive [2]. Precision is also known as positive predictive value, and recall is also known as sensitivity in diagnostic binary classification.

The F_1 score, as described earlier, is the <u>harmonic mean</u> of the precision and recall. It thus symmetrically represents both precision and recall in one metric. The more generic score applies additional weights, valuing one of precision or recall more than the other. The highest possible value of an F-score is 1.0, indicating perfect precision and recall, and the lowest possible value is 0, if either precision or recall is zero.

F_β-score is a general F-score which uses a positive real factor β, where β is chosen such that recall is considered β times as important as precision:

$$F_\beta = \left(1+\beta^2\right).\frac{Precision * Recall}{\left(\beta^2 * Precision\right) + Recall}$$

In terms of Type I and Type II errors, this becomes:

$$F_\beta = \frac{\left(1+\beta^2\right)TruePositive}{\left(1+\beta^2\right)*TruePositive+\beta^2*FalseNegative+FalsePositive}$$

7.7.4 Matthews Correlation Coefficient (MCC)

Matthews Correlation Coefficient (MCC) is a statistical tool used for model evaluation. Its job is to gauge or measure the difference between the predicted values and actual values and is equivalent to chi-square statistics for a 2 × 2 contingency table. MCC is an alternative measure unaffected by the unbalanced datasets issue, the Matthews correlation coefficient is a contingency matrix method of calculating the Pearson product-moment correlation coefficient between actual and predicted values. In terms of the entries of confusion matrix, MCC reads as follows: [3]

$$MCC = \frac{TP*TN - FP*FN}{\sqrt{(TP+FP)*(TP+FN)*(TN+FP)*(TN+FN)}}$$

Note that −1 is the worst and +1 is the best value of MCC.

MCC is the only binary classification rate that generates a high score only if the binary predictor was able to correctly predict the majority of positive data instances and the majority of negative data instances. [80, 97] It ranges in the interval [−1,+1], with extreme values −1 and +1 reached in case of perfect misclassification and perfect classification, respectively, while MCC = 0 is the expected value for the coin tossing classifier. A potential problem with MCC lies in the fact that MCC is undefined when a whole row or column of confusion matrix is zero.

7.7.5 RECEIVER OPERATING CHARACTERISTICS (ROC)

Receiver operating characteristic (ROC) curves are a popular method of summarizing the performance of classifiers. An ROC curve (receiver operating characteristic curve) is a graph showing the performance of a classification model at all classification thresholds [4]. This curve plots two parameters: True Positive Rate and False Positive Rate.

The ROC curve describes the separability of the distributions of predictions from a two-class classifier. There are a variety of situations in which an analyst seeks to aggregate multiple ROC curves into a single representative example. Several methods of doing so are available. However, there is a degree of subtlety that is often overlooked when selecting the appropriate one. An important component of this relates to the interpretation of the decision process for which the classifier will be used.

7.7.6 AREA UNDER THE ROC CURVE (AUC)

A quantity widely used to summarize an important element of the information portrayed by an ROC curve is the area under the curve (AUC). The AUC can be interpreted as an average true positive rate, regarding all values of the false positive rate as equally likely (Figure 7.4). Alternatively, it can be seen as the probability that the classifier will allocate a higher score to a randomly chosen instance

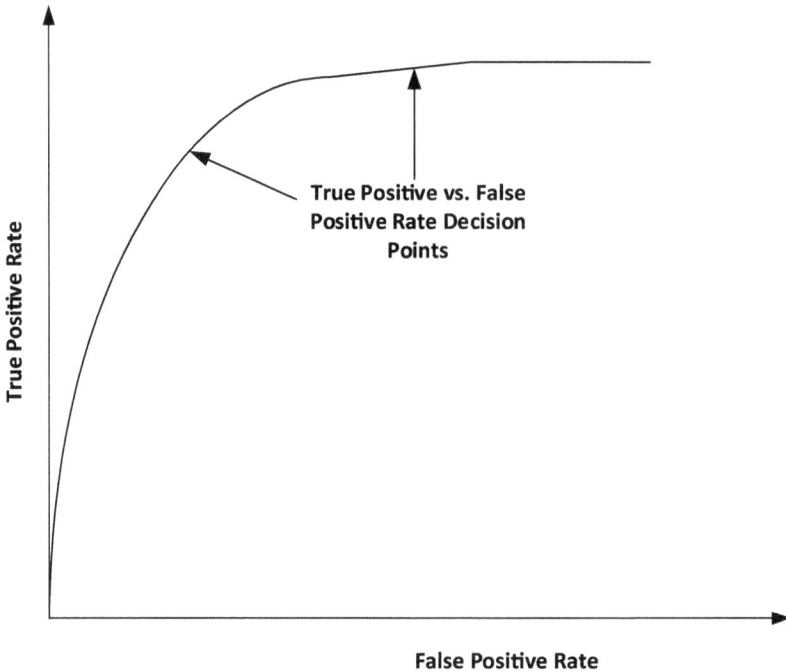

FIGURE 7.4   ROC Curve.

from P than it will to a randomly and independently chosen instance from N. Clearly, a higher AUC is desirable. However, classifiers should not be compared solely on their AUCs, as a higher AUC does not imply an everywhere dominating ROC curve. As a scalar measure of performance, a lot of valuable information contained in the ROC curve is lost, such as performance in specific regions of ROC space that may be of interest to the researcher. Moreover, there exists a fundamental incoherence in the use of AUC to compare different classifiers for an alternative interpretation of AUC.

7.7.7 EXAMPLES

It should now be clear that two fundamental considerations—the operational intention and the evaluation objective—should inform the choice of averaging method. The primary consideration relates to whether a classifier is intended to operate on all datasets using a common, fixed threshold. This dictates whether averaging should take place based on such fixed thresholds (i.e., pooling or threshold averaging) or instead in ROC space. If ROC space averaging is appropriate, a further consideration guides the choice of direction along which ROC points should be averaged. An average ROC curve can convey the performance of a classifier with respect to fixed false positive rates, true positive rates, error rate ratios, or misclassification cost ratios. Crucially, the quantity remaining fixed should be informed by the objectives of the evaluation and should be clearly stated in the presentation of results.

7.7.8 USE OF ROC CURVE

The naïve use of ROC curve combination methods is fraught with risk. We have summarized the common methods and pointed toward their interpretation in the operational context. Fundamentally, the appropriate choice of combination method must correspond to the operational approach for the classifier. We have highlighted that in the context of classifier design and development, valuable information may be lost if due care and consideration are not paid when presenting and assessing results.

7.8 SUMMARY

We have discussed the performance analysis of artificial intelligence. To evaluate performance, we need data acquisition, data digitization, data filtering (clearing) and preprocessing, feature extraction, dimensionality reduction, and data classification. We have explained that dimensionality reduction is an essential process because if the features are not relevant for classification, we need to eliminate them, thereby saving processing time and reducing complexity. However, confusion matrix, macro- and micro-averaging, F_β-Score, MCC, ROC, and AUC are used to evaluate the classifier performances. It implies that there is no single method to provide the conclusion because everything depends on what the objectives of the applications are for which the classifiers are used. We may use different algorithms for comparing the results of the classifiers to find out which one meets the objectives.

REFERENCES

[1] https://www.educative.io/answers/what-is-the-difference-between-micro-and-macro-averaging.
[2] https://en.wikipedia.org/wiki/F-score.
[3] Chico, D. and Jurman, G., "The Advantages of the Matthews Correlation Coefficient (MCC) Over F1 Score and Accuracy in Binary Classification Evaluation," BMC Genomics, 2020, https://doi.org/10.1186/s12864-019-6413-7.
[4] Poole, D. L. and Mackworth, A. K., "Artificial Intelligence: Foundation of Computational Agents," Second Edition, 2017, Cambridge University Press.

8 Unsupervised Machine Learning

Unsupervised learning is a branch of machine learning that learns from data that do not have any label. It relates to using and identifying patterns in the data for tasks such as data compression or generative models. Recently, there has been a rising trend of employing unsupervised machine learning using unstructured raw network data to improve network performance and provide services, such as traffic engineering, anomaly detection, Internet traffic classification, and quality of service optimization. The growing interest in applying unsupervised learning techniques in networking stems from their great success in other fields, such as computer vision, natural language processing, speech recognition, and optimal control (e.g., for developing autonomous self-driving cars). In addition, unsupervised learning can unconstrain us from the need for labeled data and manual handcrafted feature engineering, thereby facilitating flexible, general, and automated methods of machine learning. These machine learning algorithms discover hidden patterns or data similarity groupings without the need for any a priori knowledge or any need for human intervention and then generate imaginative content from it. Usually, dimensions reduction, visualization, associated rule learning, and clustering algorithms are used.

To develop a reliable data-driven network, data quality must be taken care before subjecting it to an appropriate unsupervised ML/DL [1]. Unsupervised ML/DL techniques facilitate the analysis of raw datasets, thereby helping in generating analytic insights from unlabeled data. Recent advances in hierarchical learning, clustering algorithms, factor analysis, latent models, and outlier detection have helped significantly advance the state of the art in unsupervised ML techniques. This significantly advanced the ML/DL state of the art by facilitating the processing of raw data without requiring careful engineering and domain expertise for feature crafting. Deep learning is a class of machine learning, where hierarchical architectures are used for unsupervised feature learning, and these learned features are then used for classification and other related tasks. The versatility of DL and distributed ML can be seen in the diversity of their applications that range from self-driving cars to the reconstruction of brain circuits. Unsupervised learning is also often used in conjunction with supervised learning in semi-supervised learning setting to preprocess the data before analysis and thereby help in crafting a good feature representation and in finding patterns and structures in unlabeled data.

The rapid advances in deep neural networks, the democratization of enormous computing capabilities through cloud computing and distributed computing, and the ability to store and process large swathes of data have motivated a surging interest in applying unsupervised ML techniques in the networking field. The field of networking also appears to be well suited to, and amenable to applications of unsupervised ML techniques, due to the largely distributed decision-making nature of its protocols, the availability of large amounts of network data, and the urgent need for intelligent/cognitive networking.

Networks these days have evolved to be very complex, and they incorporate multiple physical paths for redundancy and utilize complex routing methodologies to direct the traffic. The application traffic does not always take the optimal path we would expect, leading to unexpected and inefficient routing performance. To tame such complexity, unsupervised ML techniques can autonomously self-organize the network considering a number of factors such as real-time network congestion statistics as well as application QoS requirements.

DOI: 10.1201/9781003499466-8

8.1 HIERARCHICAL CLUSTERING

Hierarchical clustering is a popular method for grouping objects. It creates groups so that objects within a group are like each other and different from objects in other groups. Clusters are visually represented in a hierarchical tree called a dendrogram. A well-known strategy in data mining and statistical analysis in which data is clustered into a hierarchy of clusters using an agglomerative (bottom-up) or a divisive (top-down) approach. Almost all hierarchical clustering algorithms are unsupervised and deterministic. The primary advantage of hierarchical clustering over unsupervised K-means and EM algorithms is that it does not require the number of clusters to be specified beforehand. However, this advantage comes at the cost of computational efficiency. Common hierarchical clustering algorithms have at least quadratic computational complexity compared to the linear complexity of K-means and EM algorithms. Hierarchical clustering methods have a pitfall: these methods fail to accurately classify messy high-dimensional data as its heuristic may fail due to the structural imperfections of empirical data. Furthermore, the computational complexity of the common agglomerative hierarchical algorithms is NP-hard. Self-Organizing Maps (SOM) is a modern approach that can overcome the shortcomings of hierarchical models.

8.2 BAYESIAN CLUSTERING

In a Bayesian formulation of a clustering procedure, the partition of items into subsets becomes a parameter of a probability model for the data, subject to prior assumptions, and inference about the clustering derives from properties of the posterior distribution. Basically, it is a probabilistic clustering strategy where the posterior distribution of the data is learned based on a prior probability distribution. Bayesian clustering is divided into two major categories, namely parametric and non-parametric. The major difference between parametric and non-parametric techniques is the dimensionality of parameter space: if there are finite dimensions in the parameter space, the underlying technique is called Bayesian parametric; otherwise, the underlying technique is called Bayesian non-parametric. A major pitfall with the Bayesian clustering approach is that the choice of the wrong prior probability distributions can distort the projection of the data. Bayesian non-parametric clustering can be used to determine the network application type.

8.3 PARTITIONAL CLUSTERING

Partitional clustering decomposes a dataset into a set of disjoint clusters. Given a dataset of N points, a partitioning method constructs $K(N \geq K)$ partitions of the data, with each partition representing a cluster. Partitional clustering corresponds to a special class of clustering algorithms that decompose data into a set of disjoint clusters. Partitional clustering is further classified into K-means clustering and mixture models.

8.3.1 K-MEANS CLUSTERING

K-means clustering is an unsupervised learning algorithm. There is no labeled data for this clustering, unlike in supervised learning. K-means performs the division of objects into clusters that share similarities and are dissimilar to the objects belonging to another cluster. The term "K" is a number. K-means clustering is a simple, yet widely used, approach for classification. It takes a statistical vector as an input to deduce classification models or classifiers. K-means clustering tends to distribute m observations into n clusters, where each observation belongs to the nearest cluster. The membership of observation to a cluster is determined using the cluster mean. K-means clustering is used in numerous applications in the domains of network analysis and traffic classification. K-means clustering has been used in conjunction with supervised Iterative Dichotomiser 3 (ID3) decision tree learning models to detect anomalies in a network. Note that ID3 decision tree is an iterative supervised decision tree algorithm based on the concept learning system. K-means clustering provided excellent results when used in traffic classification. K-means clustering performs well in traffic classification with an accuracy of 90% [1].

K-means clustering is also used in the domain of network security and intrusion detection. The K-means algorithm can be used for intrusion detection. Experimental results on a subset of KDD-99 dataset show that the detection rate stays above 96% while the false alarm rate stays below 4%. Results and analysis of experiments on K-means algorithm have demonstrated a better ability to search clusters globally [1]. Another variation of K-means is known as K-medoids, in which rather than taking the mean of the clusters, the most centrally located data point of a cluster is considered as the reference point of the corresponding cluster.

Like K-means, K-medoids is a related algorithm that partitions data into *k* distinct clusters, by finding medoids that minimize the sum of dissimilarities between points in the data and their nearest medoid. The medoid of a set is a member of that set whose average dissimilarity with the other members of the set is the smallest. Few of the applications of K-medoids have been used in the spectrum of anomaly detection.

8.3.2 MIXTURE MODELS

Mixture models are powerful probabilistic models for univariate and multivariate data. Mixture models are used to make statistical inferences and deductions about the properties of the sub-populations given only observations on the pooled population [1]. They have also used to statistically model data in the domains of pattern recognition and computer vision. Finite mixtures, which are a basic type of mixture model, naturally model observations that are produced by a set of alternative random sources. Inferring and deducing different parameters from these sources based on their respective observations lead to clustering of the set of observations. This approach to clustering tackles drawbacks of heuristic-based clustering methods, and hence it is proven to be an efficient method for node classification in any large-scale network and has shown to yield effective results compared to techniques commonly used. For instance, K-means and hierarchical agglomerative methods rely on supervised design decisions, such as the number of clusters or validity of models [1]. Moreover, combining the expectation-maximization (EM) algorithm with mixture models produces remarkable results in deciphering the structure and topology of the vertices connected through a multidimensional network. It has been seen that Gaussian mixture model (GMM) outperforms the signature-based anomaly detection in network traffic data.

8.4   APPLICATIONS OF CLUSTERING IN NETWORKS

Network nodes based on this similarity value are grouped into the so-called *clusters*. In other words, each cluster contains elements that share common properties and characteristics. The collection of all the clusters composes a *clustering*. Clustering can be found in mostly all unsupervised learning problems, and there are diverse applications of clustering in the domain of computer networks. Two major networking applications, where significant use of clustering can be seen, are intrusion detection and Internet traffic classification. Preprocessing the data using Genetic Algorithm (GA) and then combining with the hierarchical clustering approach is called Balanced Iterative Reducing using Clustering Hierarchies (BIRCH) to provide an efficient classifier based on Support Vector Machine (SVM). This hierarchical clustering approach stores abstracted data points instead of the whole dataset, thus giving more accurate and quick classification compared to all past methods, producing better results in detecting anomalies. The use of grid-based and density-based clustering for anomaly and intrusion detection using unsupervised learning can also be used. The k-shape clustering scheme can be used for analyzing spatiotemporal heterogeneity in mobile usage. Basically, a scalable parallel framework for clustering large datasets with high dimensions is proposed and then improved by inculcating frequency pattern trees.

8.5   LATENT VARIABLE MODEL

Latent variable models involve a set of observable variables $A\{x_1, x_2, \ldots, x_k\}$ and a latent (unobservable) variable θ which may be either unidimensional (i.e., scalar) or vector valued of dimension $d \leq k$. So a latent variable model is a statistical model that relates the manifest variables with a set of latent or hidden variables. Latent variable model allows us to express relatively complex distributions in terms of tractable joint distributions over an expanded variable space [1]. Underlying variables of a process are represented in higher dimensional space using a fixed transformation, and stochastic variations are known as latent variable models where the distribution in higher dimension is due to small number of hidden variables acting in a combination. These models are used for data visualization, dimensionality reduction, optimization, distribution learning, blind signal separation and factor analysis. Next, we also have discussion on various latent variable models, namely mixture distribution, factor analysis, blind signal separation, non-negative matrix factorization, Bayesian networks and probabilistic graph models (PGM), hidden Markov model (HMM), and nonlinear dimensionality reduction techniques (which further includes generative topographic mapping, multidimensional scaling, principal curves, ISOMAP, locally linear embedding, and t-distributed stochastic neighbor embedding).

8.5.1   MIXTURE DISTRIBUTION

A mixture distribution is one that is written as a convex combination of other distributions. In fact, mixture distribution is an important latent variable model that is used for estimating the underlying density function. Mixture distribution provides a general framework for density estimation by using simpler parametric distributions. Expectation maximization (EM) algorithm is used for estimating the mixture distribution model through maximization of the log-likelihood of the mixture distribution model [1].

8.5.2   FACTOR ANALYSIS

Factor analysis is a technique that is used to reduce many variables into fewer numbers of factors. This technique extracts maximum common variance from all variables and puts them into a common score. As an index of all variables, we can use this score for further analysis. Note that factor analysis is another important type of latent variable model, which is also a density estimation model. It has been used quite often in collaborative filtering and dimensionality reduction. It is different from other latent variable models in terms of the allowed variance for different dimensions as most latent variable models for dimensionality reduction in conventional settings use a fixed variance Gaussian noise model. In the factor analysis model, latent variables have diagonal covariance rather than isotropic covariance.

8.5.3   BLIND SIGNAL SEPARATION

Blind Signal Separation (BSS) refers to a problem where the sources and the mixing matrix are indistinct and only observation signals are available for the separation procedure. The objective is to separate unknown and independent sources using observation signals. BSS, also referred to as Blind Source Separation, is the identification and separation of independent source signals from mixed input signals without or very little information about the mixing process. The most common techniques employed for BSS are principal component analysis (PCA) and independent component analysis (ICA).

8.5.3.1   Principal Component Analysis

Principal component analysis (PCA) is a dimensionality-reduction method that is often used to reduce the dimensionality of large datasets by transforming a large set of variables into a smaller one that still contains most of the information in the large set. It is a statistical procedure that utilizes orthogonal transformation on the data to convert n number of possibly correlated variables into lesser k number of uncorrelated variables called principal components. Principal components are arranged in the descending order of their variability, first one catering to the most variable and the last one to the least. Being a primary technique for exploratory data analysis, PCA takes a cloud of data in n dimensions and rotates it such that maximum variability in the data is visible. Using this technique, it brings out the strong patterns in the dataset so that these patterns are more recognizable, thereby making the data easier to explore and visualize. PCA has primarily been used for dimensionality reduction in which input data of n dimensions is reduced to k dimensions without losing critical information in the data. The choice of the number of principal components is a question of the design decision. Much research has been conducted on selecting the number of components such as cross-validation approximations. Optimally, k is chosen such that the ratio of the average squared projection error to the total variation in the data is less than or equal to 1% by which 99% of the variance is retained in the k principal components [1]. But, depending on the application domain, different designs can increase/decrease the ratio while maximizing the required output. Commonly, many features of a dataset are often highly correlated. Hence, PCA results in retaining 99% of the variance while significantly reducing the data dimensions.

8.5.3.2 Independent Component Analysis

Independent Component Analysis (ICA) is a technique that allows the separation of a mixture of signals into their different sources by assuming non-Gaussian signal distribution. The ICA extracts the sources by exploring the independence underlying the measured data. Note that ICA is another technique for BSS that focuses on separating multivariate input data into additive components with the underlying assumption that the components are non-Gaussian and statistically independent. The most common example to understand ICA is the cocktail party problem in which there are n people talking simultaneously in a room and one tries to listen to a single voice. ICA separates source signals from input mixed signal by either minimizing the statistical dependence or maximizing the non-Gaussian property among the components in the input signals by keeping the underlying assumptions valid. Statistically, ICA can be seen as the extension of PCA, whereas PCA tries to maximize the second moment (variance) of data, hence relying heavily on Gaussian features; on the other hand, ICA exploits inherently non-Gaussian features of the data and tries to maximize the fourth moment of linear combination of inputs to extract non-normal source components in the data.

8.5.4 Non-Negative Matrix Factorization

Non-negative matrix factorization (NMF or NNMF), also non-negative matrix approximation, is a group of algorithms in multivariate analysis and linear algebra where a matrix V is factorized into (usually) two matrices W and H, with the property that all three matrices have no negative elements. So NMF is a technique to factorize a large matrix into two or more smaller matrices with no negative values, that is when multiplied, it reconstructs the approximate original matrix. NMF is a novel method for decomposing multivariate data making it easy and straightforward for exploratory analysis. By NMF, hidden patterns and intrinsic features within the data can be identified by decomposing them into smaller chunks, enhancing the interpretability of data for analysis, with positivity constraints. However, there exist many classes of algorithms [1] for NMF having different generalization properties, for example, two of them are analyzed in [101], one of which minimizes the least square error while the other focuses on the Kullback-Leibler divergence keeping algorithm convergence intact.

8.5.5 Hidden Markov Models

Hidden Markov models (HMMs) are sequence models. That is, given a sequence of inputs, such as words, a HMM will compute a sequence of outputs of the same length. A HMM is a graph where nodes are probability distributions over labels, and edges give the probability of transitioning from one node to the other. HMMs are stochastic models of great utility, especially in domains where we wish to analyze temporal or dynamic processes, such as speech recognition, primary users (PU) arrival pattern in cognitive radio networks (CRNs), and others. HMMs are highly relevant to CRNs since many environmental parameters in CRNs are not directly observable. A HMM-based approach can analytically model a Markovian stochastic process in which we do not have access to the actual states, which are assumed to be unobserved or hidden, instead, we can observe a state that is stochastically dependent on the hidden state. It is for this reason that a HMM is defined to be a doubly stochastic process.

8.5.6 Bayesian Networks and Probabilistic Graph Models

Bayesian Networks and Probabilistic Graph Models are probabilistic graphical models for representing knowledge about an uncertain domain, where each node corresponds to a random variable and each edge represents the conditional probability for the corresponding random variables. A Bayesian network (also known as a Bayes network, Bayes net, belief network, or decision network) is a probabilistic graphical model that represents a set of variables and their conditional dependencies via a directed acyclic graph (DAG). It is one of several forms of causal notation. In Bayesian learning, we try to find the posterior probability distributions for all parameter settings; in this setup, we ensure that we have a posterior probability for every possible parameter setting. It is computationally expensive, but we can use complicated models with a small dataset and still avoid overfitting. Posterior probabilities are calculated by dividing the product of sampling distribution and prior distribution by marginal likelihood. In simple words, posterior probabilities are calculated using Bayes theorem. The basis of reinforcement learning was also derived by using Bayes theorem [1]. Since Bayesian learning is computationally expensive, a new research trend is approximate Bayesian learning. With the emergence of Bayesian deep learning framework, the deployment of Bayes learning-based solution is increasing rapidly. Probabilistic graph modeling is a concept associated with Bayesian learning. A model representing the probabilistic relationship between random variables through a graph is known as a probabilistic graph model (PGM). Nodes and edges in the graph represent a random variable and their probabilistic dependence, respectively. PGMs are of two types: directed PGM and undirected PGM. Bayes networks also fall in the regime of directed PGM. PGM is used in many important areas such as computer vision, speech processing, and communication systems. Bayesian learning combined with PGM, and latent variable models, forms a probabilistic framework where deep learning is used as a substrate for making improved learning architecture for recommender systems, topic modeling, and control systems.

8.5.7 Applications of Latent Variable Models in Networks

Latent variables are often used in structural equation modeling (SEM), which is a statistical technique used to test complex theoretical models. SEM allows researchers to test the relationships between latent variables and observed variables and to estimate the strength of those relationships. Latent variables methods have applied latent structure on email corpus to find interpretable latent structure as well as evaluating its predictive accuracy on missing data task. A dynamic latent model has been applied for a social network. BSS is a set of statistical algorithms that are widely used in different application domains to perform different tasks such as dimensionality reduction, correlating and mapping features, and others. PCA is employed for Internet traffic classification to separate different types of flows in a network packet stream. Similarly, in a semi-supervised approach, PCA is used for feature learning and an SVM classifier for intrusion detection in an autonomous network system. Another approach for detecting anomalies and intrusions proposed uses NMF to factorize different flow features and cluster them accordingly. Furthermore, ICA has been widely used in telecommunication networks to separate mixed and noisy source signals for efficient service. For example, a variant of

ICA called Efficient Fast ICA (EF-ICA) has been used for detecting and estimating the symbol signals from the mixed code-division multiple access (CDMA) signals received from the source endpoint. In other literature, PCA uses a probabilistic approach to find the degree of confidence in detecting an anomaly in wireless networks. Furthermore, PCA is also chosen as a method of clustering and designing Wireless Sensor Networks (WSNs) with multiple sink nodes [1].

Bayesian learning has been applied for classifying Internet traffic, where Internet traffic is classified based on the posterior probability distributions. For early traffic identification in campus network, real discretized conditional probability has been used to construct a Bayesian classifier. Host-level intrusion detection and anomalies classification in border gateway protocol (BGP) using Bayesian networks has been used. Internet threat detection and outlier detection using Bayesian belief networks are also described [1]. A new approach toward outlier detection using Bayesian belief networks is described in [1]. Application of Bayesian networks in MIMO systems has been explored. Location estimation using Bayesian network in local area network (LAN) is discussed. Similarly, both Bayes theory and PGM are used in Low-Density Parity Check (LDPC) and Turbo codes, which are the fundamental components of information coding theory.

8.6   DIMENSIONALITY REDUCTION

Representing data in fewer dimensions is another well-established task of unsupervised learning. Real world data often have high dimensions—in many datasets, these dimensions can run into thousands, even millions, of potentially correlated dimensions. However, it is observed that the intrinsic dimensionality (governing parameters) of the data is less than the total number of dimensions. To find the essential pattern of the underlying data by extracting intrinsic dimensions, it is necessary that the real essence is not lost, for example, it may be the case that a phenomenon is observable only in higher-dimensional data and is suppressed in lower dimensions, such phenomena are said to suffer from the curse of dimensionality. While dimensionality reduction is sometimes used interchangeably with feature selection, a subtle difference exists between the two.

Feature selection is traditionally performed as a supervised task with a domain expert helping in handcrafting a set of critical features of the data. Such an approach generally can perform well but is not scalable and prone to judgment bias. Dimensionality reduction, on the other hand, is more generally an unsupervised task, where instead of choosing a subset of features, it creates new features (dimensions) as a function of all features. Said differently, feature selection considers supervised data labels while dimensionality reduction focuses on the data points and their distributions in an N-dimensional space. There exist different techniques for reducing data dimensions including projection of higher dimensional points onto lower dimensions, independent representation, and sparse representation, which should be capable of reconstructing the approximate data. Dimensionality reduction is useful for data modeling, compression, and visualization.

By creating representative functional dimensions of the data and eliminating redundant ones, it becomes easier to visualize and form a learning model. Independent representation tries to disconnect the source of variation underlying the data distribution such that the dimensions of the representation are statistically independent. Sparse representation technique represents the data vectors in linear combinations of small basis vectors. It is worth noting here that many of the latent variable models (e.g., PCA, ICA, factor analysis) also function as techniques for dimensionality reduction. In addition to techniques such as PCA and ICA—which infer the latent inherent structure of the data through a linear projection of the data—several nonlinear dimensionality reduction techniques have also been developed. Linear dimensionality reduction techniques are useful in many settings, but these methods may miss important nonlinear structure in the data due to their subspace assumption, which posits that the high-dimensional data points lie on a linear subspace (e.g., on a 2D or 3D plane). Such an assumption fails in high dimensions when data points are random but highly correlated with neighbors. In such environments nonlinear dimensionality reductions through manifold learning techniques—which can be construed as an attempt to generalize linear frameworks like

PCA so that nonlinear structure in data can also be recognized—become desirable. Even though some supervised variants also exist, manifold learning is mostly performed in an unsupervised fashion using the nonlinear manifold substructure learned from the high-dimensional structure of the data from the data itself without the use of any predetermined classifier or labeled data. Some nonlinear dimensionality reduction (manifold learning) techniques are described in the next section.

8.6.1 ISOMAP

ISOMAP is used for computing a quasi-isometric, low-dimensional embedding of a set of high-dimensional data points. The algorithm provides a simple method for estimating the intrinsic geometry of a data manifold based on a rough estimate of each data point's neighbors on the manifold. That is, ISOMAP is a nonlinear dimensionality reduction technique that finds the underlying low-dimensional geometric information about a dataset. Also, algorithmic features of PCA and multidimensional scaling (MDS) are combined to learn the low-dimensional nonlinear manifold structure in the data. ISOMAP uses geodesic distance along the shortest path to calculate the low-dimension representation shortest path, which can be computed using Dijkstra's algorithm. MDS is a simple distance matrix. In PCS, a dataset where each row is an observation, and each column is a feature.

8.6.2 GENERATIVE TOPOGRAPHIC MAPPING

Generative topographic mapping (GTM) represents the nonlinear latent variable mapping from continuous low-dimensional distributions embedded in high-dimensional spaces. Data space in GTM is represented as reference vectors, and these vectors are a projection of latent points in data space. GTM is a map of a region that shows changes in elevation, usually with contour lines indicating different fixed elevations. It is a probabilistic variant of self-organizing maps (SOM) and works by calculating the Euclidean distance between data points. GTM optimizes the log-likelihood function, and the resulting probability defines the density in data space.

8.6.3 LOCALLY LINEAR EMBEDDING

Locally Linear Embedding (LLE) is a dimensionality reduction technique used in machine learning and data analysis. It focuses on preserving local relationships between data points when mapping high-dimensional data to a lower-dimensional space. LLE is an unsupervised nonlinear dimensionality reduction algorithm. LLE represents data in lower dimensions yet preserving the higher-dimensional embedding. LLE depicts data in a single global coordinate of lower-dimensional mapping of input data. LLE is used to visualize multidimensional manifolds and feature extraction.

8.6.4 PRINCIPAL CURVE

Principal curves are smooth one-dimensional curves that pass through the middle of a p-dimensional dataset, providing a nonlinear summary of the data. It is a nonlinear dataset summarizing technique, where non-parametric curves pass through the middle of a p-dimensional (multidimensional) dataset providing the summary of the dataset. These smooth curves minimize the average squared orthogonal distance between data points; this process also resembles the maximum likelihood for nonlinear regression in the presence of Gaussian noise.

8.6.5 NONLINEAR MULTI-DIMENSIONAL SCALING

Non-metric Multi-Dimensional Scaling (NMDS) is a distance-based ordination technique. The data matrix (n sample units $\times p$ species) is converted into an $n \times n$ distance matrix (or, more generally, a dissimilarity matrix), and the ordination is based upon the distance matrix. NMDS is a nonlinear latent variable representation scheme. It works as an alternative scheme for factor analysis. In factor analysis, a multivariate normal distribution is assumed, and similarities between different objects are expressed as a correlation matrix. Whereas NMDS does not impose such a condition, and it is designed to reach the optimal low-dimensional configuration where similarities and dissimilarities among matrices can be observed. NMDS is also used in data visualization and mining tools for depicting the multidimensional data in three dimensions based on the similarities in the distance matrix.

8.6.6 T-Distributed Stochastic Neighbor Embedding

The T-distributed stochastic neighbor embedding (t-SNE) describes the standardized distances of sample means to the population mean when the population standard deviation is not known, and the observations come from a normally distributed population. t-SNE is another nonlinear dimensionality reduction scheme. It is used to represent high-dimensional data in 2D or 3D dimensions. t-SNE constructs a probability distribution in high-dimensional space and constructs a similar distribution in lower dimensions and minimizes the Kullbackâ-ĂŞLeibler (KL) divergence between two distributions (which is a useful way to measure the difference between two probability distributions).

8.7 OUTLIER DETECTION

Outlier detection is the process of identifying data points that are significantly different from the rest. The three main outlier detection methods in data mining are statistical, proximity-based, and Model-Based. Outlier detection is an important application of unsupervised learning. A sample point that is distant from other samples is called an outlier. An outlier may occur due to noise, measurement error, heavy tail distributions, and a mixture of two distributions. There are two popular underlying techniques for unsupervised outlier detection upon which many algorithms are designed, namely the nearest neighbor-based technique and clustering-based method.

8.7.1 Nearest Neighbor Anomaly Detection

Nearest neighbor anomaly detection techniques assume that normal data instances occur in dense neighborhoods while anomalies occur far from their closet neighbors and require a distance or similarity measure defined between two data instances. Distance or similarity can be computed in different ways. The nearest neighbor method works on estimating the Euclidean distances or average distance of every sample from all other samples in the dataset. There are many algorithms based on nearest neighbor-based techniques, with the most famous extension of the nearest neighbor being a k-nearest neighbor technique in which only k-nearest neighbors participate in the outlier detection.

8.7.2 Local Outlier Factor

The Local Outlier Factor (LOF) algorithm is an unsupervised anomaly detection method which computes the local density deviation of a given data point with respect to its neighbors. It considers as outliers the samples that have a substantially lower density than their neighbors. It works as an extension of the k-nearest neighbor algorithm.

8.7.3 Connectivity-Based Outlier Factors

Connectivity-based outlier factors (COF) compute the connectivity-based outlier factor for observations, being the comparison of chaining-distances between observation subject to outlier scoring and neighboring observations. The COF function is useful for outlier detection in clustering and other multidimensional domains.

8.7.4 INFLUENCED OUTLIERNESS

Influenced Outlierness (INFLO) computes the influenced outlierness score for observations, being the comparison of density in neighborhood of observation subject to outlier scoring and density in the reverse neighborhood. The INFLO function is useful for outlier detection in clustering and other multidimensional domains.

8.7.5 LOCAL OUTLIER PROBABILITY MODELS

Local Outlier Probabilities (LoOP) is a local density-based outlier detection approach which provides outlier scores in the range of [0,1] that are directly interpretable as the probability of a sample being an outlier.

8.8 KEY CHARACTERISTIC OF UNSUPERVISED LEARNING

The key characteristics of unsupervised learning techniques are summarized here:

- Hierarchical learning techniques are the most popular schemes in literature for feature detection and extraction.
- Learning the joint distribution of a complex distribution over an expanded variable space is a difficult task. Latent variable models have been the recommended and well-established schemes in literature for this problem. These models are also used for dimensionality reduction and better representation of data.
- Visualization of unlabeled multidimensional data is another unsupervised task. In this research, we have explored dimensionality reduction as an underlying scheme for developing better multidimensional data visualization tools.

8.9 APPLICATIONS OF UNSUPERVISED LEARNING IN NETWORKING

We highlight the broad spectrum of applications in networking and emphasize the importance of ML/DL-based techniques, rather than classical hard-coded statistical methods, for achieving more efficiency, adaptability, and performance enhancement. Some major usages of Unsupervised Learning in Networking are listed as follows:

- Internet Traffic Classification
- Anomaly/Intrusion Detection
- Network Operations, Optimization, and Analytics
  - Quality-of-Service/Quality-of-Experience
  - Transmission Control Protocol Optimization
- Dimensionality Reduction and Virtualization
- Emerging Networking Applications
  - Software Defined Network
  - Internet-of-Things
  - Transmission Control Protocol Optimization

Some recommendations related to unsupervised learning in networking applications are summarized as follows:

- A recommended and well-studied method for unsupervised Internet traffic classification in literature is data clustering combined with the latent representation learning on traffic feature set by using autoencoders. Min-max ensemble learning will help to increase the efficiency of unsupervised learning if required.

- Semi-supervised learning is also an appropriate method for Internet traffic classification given some labeled traffic data and channel characteristics are available for initial model training.
- Application of generative models and transfer learning for the Internet traffic classification has not been explored properly in literature and can be a potential research direction.
- The overwhelming growth in network traffic and expected surge in traffic with the evolution of 5G and IoT also elevates the level of threat and anomalies in network traffic. To deal with these anomalies in Internet traffic, data clustering, PCA, SOM, and ART are well-explored unsupervised learning techniques in the literature. Self-taught learning has also been explored as a potential solution for anomaly detection and remains a possible research direction for future research in anomaly detection in network traffic.
- Current state of the art in dimensionality reduction in network traffic is based on PCA and multidimensional scaling. Autoencoders, t-SNE, and manifold learning are potential areas of research in terms of dimensionality reduction and visualization.

8.10 FUTURE NETWORK: RESEARCH CHALLENGES AND OPPORTUNITIES

We provide a discussion on some open directions for future work and the relevant opportunities in applying unsupervised ML/DL in the field of networking.

8.10.1 SIMPLIFIED NETWORK MANAGEMENT

While new network architectures such as software defined network (SDN) have been proposed in recent years to simplify network management, network operators are still expected to know too much and to correlate between what they know about how their network is designed with the current network's condition through their monitoring sources. Operators who manage these requirements by wrestling with complexity manually will welcome any respite that they can get from (semi-)automated unsupervised machine learning. ML/DL to become pervasive in networking, the "semantic gap"—which refers to the key challenge of transferring ML results into actionable insights and reports for the network operator—must be overcome. This can facilitate a shift from a reactive interaction style for network management, where the network manager is expected to check maps and graphs when things go wrong, to a proactive one, where automated reports and notifications are created for different services and network regions. This will require a move beyond mere notifications and visualizations to more substantial synthesis through which potential sources of problems can be identified. Another example relates to making measurements more user-oriented. Most users would be more interested in QoE instead of QoS, i.e., how the current condition of the network affects their applications and services rather than just raw QoS metrics. The development of measurement objectives should be from a business-eyeball perspective—and not only through presenting statistics gathered through various tools and protocols such as traceroute, ping, border gateway protocol (BGP), and others with the burden of putting the various pieces of knowledge together being on the user.

8.10.2 SEMI-SUPERVISED LEARNING FOR COMPUTER NETWORKS

Semi-supervised learning lies between supervised and unsupervised learning. The idea behind semi-supervised learning is to improve the learning ability by using unlabeled data incorporation with a small set of labeled examples. In computer networks, semi-supervised learning is partially used in anomaly detection and traffic classification and has great potential to be used with deep unsupervised learning architectures like generative adversarial networks for improving the state of the art in anomaly detection and traffic classification. Similarly, user behavior learning for cybersecurity can also be tackled in a semi-supervised fashion. A semi-supervised learning-based anomaly detection approach is also studied. The presented approach used large amounts of unlabeled samples together with labeled samples to build a better intrusion detection classifier. A single hidden layer feed forward neural network (NN) has trained to output a fuzzy membership vector. The results show that using unlabeled samples helps significantly improve the classifier's performance.

8.10.3 Transfer Learning in Computer Networks

Transfer learning is an emerging ML/DL technique in which knowledge learned from one problem is applied to a different but related problem. Although it is often thought that for ML algorithms, the training and future data must be in the same feature space and must have the same distribution, this is not necessarily the case in many real-world applications. In such cases, it is desirable to have transfer learning or knowledge transfer between the different task domains. Transfer learning has been successfully applied in computer vision and natural language processing (NLP) applications, but its implementation for networking has not been witnessed—even though in principle, this can be useful in networking as well due to the similar nature of Internet traffic and enterprise network traffic in many respects. The transfer learning-based caching procedure for wireless networks providing backhaul offloading in 5G networks has also been investigated.

8.10.4 Federated Learning in Computer Networks

Federated learning is a collaborative ML/DL technique, which does not make use of centralized training data, and works by distributing the processing on different machines. Federated learning is considered to be the next big thing in cloud networks as they ensure the privacy of the user data and less computation on the cloud to reduce the cost and energy. System and method for network address management in the federated cloud are presented and the application of federated IoT and cloud computing for health care has been used. An end-to-end security architecture for federated cloud and IoT is investigated.

8.10.5 Generative Adversarial Networks in Computer Networks

Adversarial networks—based on generative adversarial network (GAN) training originally proposed by Goodfellow and colleagues at the University of Montreal—have recently emerged as a new technique using which machines can be trained to predict outcomes by only observing the world (without necessarily being provided labeled data). An adversarial network has two neural network (NN) models:

- A generator which is responsible for generating some type of data from some random input; and
- A discriminator, which has the task of distinguishing between input from the generator or a real dataset.

The two NNs optimize themselves together resulting in a more realistic generation of data by the generator and a better sense of what is plausible in the real world for the discriminator. A GAN for generating malware examples to attack a malware classifier and then proposes a defense against it has also been proposed. Another adversarial perturbation attack on malware classifier is proposed. The use of GANs for ML/DL in networking can improve the performance of ML/DL-based networking applications such as anomaly detection in which malicious users have an incentive to craft new attacks to avoid detection by network managers.

8.11 PITFALLS AND CAVEATS OF USING ML/DL IN NETWORKING

With the benefits and intriguing results of unsupervised learning, there also exist many shortcomings that are not addressed widely in the literature. Some potential pitfalls and caveats related to unsupervised learning are discussed next.

8.11.1 Inappropriate Technique Selection

To start with, the first potential pitfall could be the selection of technique. Different unsupervised learning and predicting techniques may have excellent results on some applications while performing poorly on others—it is important to choose the best technique for the task at hand. Another reason could be a poor selection of features or parameters on which basis predictions are made—thus parameter optimization is also important for unsupervised algorithms.

8.11.2 Lack of Interoperability of Some Supervised ML/DL Algorithms

Some unsupervised algorithms such as deep NNs operate as a black box, which makes it difficult to explain and interpret the working of such models. This makes the use of such techniques unsuitable for applications in which interpretability is important. Understandability of the semantics of the decisions made by ML/DL is especially important for the operational success of ML/DL in large-scale operational networks and its acceptance by operators, network managers, and users. But prediction accuracy and simplicity are often in conflict. As an example, the greater accuracy of NNs accrues from its complex nature in which input variables are combined in a nonlinear fashion to build a complicated hard-to-explain model; with NNs it may not be possible to get interpretability as well since they make a tradeoff in which they sacrifice interpretability to achieve high accuracy. There are various ongoing research efforts that are focused on making techniques such as NNs less opaque. Apart from the focus on NNs, there is a general interest in making AI and ML/DL more explainable and interpretable—e.g., the Defense Advanced Research Projects Agency or DARPA's explainable AI project #2 is aiming to develop explainable AI models (leveraging various design options spanning the performance-vs-explainability tradeoff space) that can explain the rationale of their decision-making so that users are able to appropriately trust these models particularly for new envisioned control applications in which optimization decisions are made autonomously by algorithms.

8.11.3 Lack of Operations Success of ML/DL Networking

In literature, researchers have noted that despite substantial academic research, and practical applications of unsupervised learning in other fields, we see that there is a dearth of practical applications of ML/DL solutions in operational networks—particular for applications such as network intrusion detection—which are challenging problems for several reasons including the following:

- Very high cost of errors
- Lack of training data
- The semantic gap between results and their operational interpretation
- Enormous variability in input data
- Fundamental difficulties in conducting sound performance evaluations

Even for other applications, the success of ML/DL and its wide adoption in practical systems at scale lags the success of ML/DL solutions in many other domains.

8.11.4 Ignoring Simple Non-ML/DL-Based Tools

One should also keep in mind a common pitfall that academic researchers may suffer from is not realizing that network operators may have simpler non-machine learning-based solutions that may work as well as naïve ML/DL-based solutions on practical settings. Failure to examine the ground realities of operational networks will undermine the effectiveness of ML/DL-based solutions. We should expect ML/DL-based solutions to augment and supplement rather than replace other non-machine-learning based solutions—at least for the foreseeable future.

8.11.5 OVERFITTING

Another potential issue with unsupervised models is overfitting; it corresponds to a model representing the noise or random error rather than learning the actual pattern in data. While commonly associated with supervised ML/DL, the problem of overfitting lurks whenever we learn from data and thus is applicable to unsupervised ML/DL as well. Ideally speaking, we expect ML/DL algorithms to provide improved performance with more data, but with increasing model complexity, performance starts to deteriorate after a certain point—although, it is possible to get poorer results empirically with increasing data when working with unoptimized out-of-the-box ML/DL algorithms. According to the Occam Razor principle, the model complexity should be commensurate with the amount of data available and, with overly complex models, the ability to predict and generalize diminishes. Two major reasons for overfitting could be the overly large size of the learning model and fewer sample data used for training purposes. Generally, data is divided into two portions (actual data and stochastic noise). Due to the unavailability of labels or related information, unsupervised learning model can overfit the data, which causes issues in testing and deployment phase. Cross-validation, regularization, and Chi-squared testing are highly recommended for designing or tweaking an unsupervised learning algorithm to avoid overfitting.

8.11.6 DATA QUALITY ISSUES

It should be noted that all ML/DL is data dependent, and the performance of ML/DL algorithms is affected largely by the nature, volume, quality, and representation of data. In the case of unsupervised ML/DL data, quality issues must be carefully considered since any problem with the data quality will seriously mar the performance of ML/DL algorithms. A potential problem is that dataset may be imbalanced if the sample size from one class is very much smaller or larger than the other classes. In such imbalanced datasets, the algorithm must be careful not to ignore the rare class by assuming it to be noise. Although imbalanced datasets are more of a nuisance for supervised learning techniques, they may also pose problems for unsupervised and semi-supervised learning techniques.

8.11.7 INACCURATE MODEL BUILDING

It is difficult to build accurate and generic models since each model is optimized for certain kind of applications. Unsupervised ML/DL models should be applied after carefully studying the application and the suitability of the algorithm in such settings. For example, we highlight certain issues related to the unsupervised task of clustering:

- Random initialization in K-means is not recommended.
- Number of clusters is not known before the clustering operation as we do not have labels.
- In the case of hierarchical clustering, we do not know when to stop, and this can increase the time complexity of the process.
- Evaluating the clustering result is very tricky since the ground truth is mostly unknown.

8.11.8 MACHINE LEARNING IN ADVERSARIAL ENVIRONMENTS

Many networking problems, such as anomaly detection, are adversarial problems in which the malicious intruder is continually trying to outwit the network administrators (and the tools used by the network administrators). In such settings, machine learning that learns from historical data may not perform due to clever crafting of attacks specifically for circumventing any schemes based on previous data. Due to these challenges, pitfalls, and weaknesses, due care must be exercised while using unsupervised and semi-supervised ML/DL. These pitfalls can be avoided in part by using various best practices, such as end-to-end learning pipeline testing, visualization of the learning algorithm, regularization, proper feature engineering, dropout, sanity checks through human inspection—whichever is appropriate for the problem's context.

8.12 SUMMARY

We have provided a comprehensive survey of machine learning tasks, latest unsupervised learning techniques, and trends, along with a detailed discussion of the applications of these techniques in networking related tasks. Despite the recent wave of success of unsupervised learning, there is a scarcity of unsupervised learning literature for computer networking applications, which this survey aims to address. Due to the versatility and evolving nature of computer networks, it was impossible to cover every application. However, an attempt has been made to cover all the major networking applications of unsupervised learning and the relevant techniques. We have also presented concise future work and open research areas in the field of networking, which is related to unsupervised learning, coupled with a brief discussion of significant pitfalls and challenges in using unsupervised machine learning in networks.

REFERENCE

[1] Usama, M. et al.," Unsupervised Machine Learning for Networking: Techniques, Applications and Research Challenges," IEEE Access, June 2019. Digital Object Identifier, https://doi.org/10.1109/ACCESS.2019.2916648.

9 Supervised Machine Learning

Supervised learning uses labeled datasets to train algorithms, and it is used to classify data or outcomes accurately based on the learning criteria learned during training. A feature is a function from examples into a value. Let us consider a set of input features $\{x\}_1^n$ and a set of target features $\{y\}_1^k$ that are given for each example. A set of test examples where only the values for the input features are also given. A supervised learning algorithm can be viewed as a function that maps a dataset D of learning samples (X, Y). A dataset D of learning samples $(x, y) \sim (X, Y)$ with independent and identically distributed (i.i.d.) random variables in a model. The formulation is done as follows: [1]

$$\text{Function} - f : X \to Y$$

$$\text{Input} - x \in X$$

$$\text{Output} - y \in Y$$

x and y depend on application and is provided by:

$$y = f(x)$$

A supervised learning algorithm can be viewed as a function that maps a dataset D of learning samples $(x, y) \sim (X, Y)$. Input function is related by $f(x \mid D)$—a random variable. So its average error over the input-space is. The expected value of the error is given by:

$$e[f] = \underset{X}{E}\, \underset{D}{E}\, \underset{Y|X}{E}\, L\left(Y, f\left(X|D\right)\right)$$

where $L(.,.)$ is a loss function.
   If

$$L(y, \hat{y}) = (y - \hat{y})^2$$

The error decomposes naturally into a sum of a bias term and a variance term. This bias-variance decomposition can be useful because it highlights a tradeoff between an error due to erroneous assumptions in the model selection/learning algorithm (the bias) and an error because only a finite set of data is available to learn that model (the parametric variance). Note that the parametric variance is also called the overfitting error. Without knowing the joint probability distribution, it is impossible to compute $e[f]$. Instead, we can compute the empirical error on a sample of data. Given n data points (x_i, y_i), the empirical error is:

$$e_S[f] = \frac{1}{n}\sum_{i=1}^{n} L\left(y_i, f\left(y_i\right)\right)$$

DOI: 10.1201/9781003499466-9

The generalization error is the difference between the error on a sample set (used for training) and the error on the underlying joint probability distribution. It is defined as:

$$G = e[f] - e_S[f]$$

The bias-variance decomposition is given by:

$$\underset{D}{E}\,\underset{Y|X}{E}\left(Y - f\left(X|D\right)\right)^2 = \sigma^2(x) + \text{bias}^2(x)$$

where

$$\text{bias}^2(x) \triangleq \left(\underset{Y|x}{E}(Y) - \underset{D_{LS}}{E} f(x|D)\right)^2$$

$$\sigma^2(x) \triangleq \underset{Y|x}{E}\left(Y - \underset{Y|x}{E}(Y)\right)^2 + \underset{D_{LS}}{E}\left(f(x|D) - \underset{D_{LS}}{E} f(x|D)\right)^2$$

$$\underset{Y|x}{E}\left(Y - \underset{Y|x}{E}(Y)\right)^2 = \text{Internal Variance}$$

$$\underset{D}{E}\left(f(x|D) - \underset{D}{E} f(x|D)\right)^2 = \text{Parametric Variance}$$

For any given model, the parametric variance goes to zero with an arbitrary large dataset by considering the strong law of convergence. A wide range of supervised learning algorithms are available, each with its strengths and weaknesses. Examples are linear regression, neural network with multiple perception, linear regression, support vector machine, Naïve Bayes, and Decision Tree, K-nearest neighbor algorithm, similarity learning, and others. Image recognition, fraud detection, spam filters, classification, and regression are examples of supervised learning. However, no unknown patterns that are not trained with could not be recognized by supervised learning.

9.1 LINEAR REGRESSION AND CLASSIFICATION

Regression algorithms provide prediction with continuous value (e.g., price, salary, and age) with labeled datasets, but classification algorithms provide prediction or classification in discrete values such as male or female, true or false, spam or not spam. We are considering a linear regression for fitting a linear function to a set of training examples in which the input and target features are numeric—Input Feature $\{X_i\}_1^n$ and Target Feature $\{Y\}$—and a linear function of the input feature is a function of the form:

$$\hat{Y}^{\bar{w}}(e) = \sum_{i=0}^{n} w_i * X_i(e)$$

where $\bar{w} = \langle w_0, w_1, \ldots, w_n \rangle$ is a tuple of weights [1]. To make w_0 not a special case, we invent a new feature, X_0, whose value is always 1. The sum-of-square error for an example sample set of E_s for target Y is:

$$error\left(E_s, \bar{w}\right) = \sum_{e \in E_s}\left[\left(Y(e) - \hat{Y}^{\bar{w}}(e)\right)\right]^2 = \sum_{e \in E_s}\left[\left(Y(e) - \sum_{i=0}^{n} w_i * X_i(e)\right)\right]^2$$

Gradient descent is an optimization algorithm which is commonly used to train machine learning models and neural networks. Training data helps these models learn over time, and the cost function within gradient descent specifically acts as a barometer, gauging its accuracy with each iteration of parameter updates. If we differentiate our error function $\frac{\partial}{\partial w_i}\left[error\left(E_s,\overline{w}\right)\right]$ to minimize error in each step, it decreases each weight in proportion to its partial derivative as follows: [1]

$$w_{i+1} = w_i - \eta * \frac{\partial}{\partial w_i}\left[error\left(E_s,\overline{w}\right)\right]$$

where η is the gradient descent step size and is also known as the learning rate. The learning rate, the feature, and the data are provided as input to the algorithm with the aim for minimizing the error (e.g., sum-of-the error).

9.2   LOGISTIC REGRESSION

Logistic regression is the statistical model (also known as logit model) and is often used for classification and predictive analytics. Logistic regression estimates the probability of an event occurring, such as voted or did not vote, based on a given dataset of independent variables. The logistic regression model transforms the linear regression function's continuous value output into categorical value output using a sigmoid function, which maps any real-valued set of independent variables input into a value between 0 and 1. This function is known as the logistic function.

The assumptions for logistic regression are as follows:

- **Independent Observations:** Each observation is independent of the other, meaning there is no correlation between any input variables.
- **Binary Dependent Variables:** It takes the assumption that the dependent variable must be binary or dichotomous, meaning it can take only two values. For more than two categories SoftMax functions are used.
- **Linearity Relationship between Independent Variables and Log Odds:** The relationship between the independent variables and the log odds of the dependent variable should be linear.
- **No Outliers:** There should be no outliers in the dataset.
- **Large Sample Size:** The sample size is sufficiently large.

Let the independent input features be: [2]

$$X \equiv \begin{bmatrix} x_{11} & \cdots & x_{1m} \\ \cdots & \cdots & \cdots \\ x_{n1} & \cdots & x_{nm} \end{bmatrix}$$

and the dependent variable is Y having only binary value, i.e., 0 or 1.

$$Y = \begin{cases} 0 & if\ Class\ 1 \\ 1 & if\ Class\ 2 \end{cases}$$

then apply the multilinear function to the input variables X.

$$z = \left(\sum_{i=1}^{n} w_i x_i \right) + b$$

Here, x_i is the *ith* observation of X, $w_i = [w_1, w_2, \ldots, w_m]$ is the weights or Coefficient, and b is the bias term, also known as intercept. This can be simply represented as the dot product of weight and bias:

$$z = w.X + b$$

whatever we have discussed previously is the linear regression.

Now we consider the sigmoid function where the input will be z, and we find the probability between 0 and 1 that is predicted by y (see Figure 9.1).

$$\sigma(z) = \frac{1}{1 - e^{-z}}$$

The sigmoid function converts the continuous variable data into probability, that is, between 0 and 1.

$$\sigma(z) \to 1 \text{ as } z \to \infty$$

$$\sigma(z) \to 0 \text{ as } z \to -\infty$$

So the sigmoid function is always bounded between 0 and 1 where the probability of being a class can be measured as:

$$P(y = 1) = \sigma(z)$$

$$P(y = 0) = 1 - \sigma(z)$$

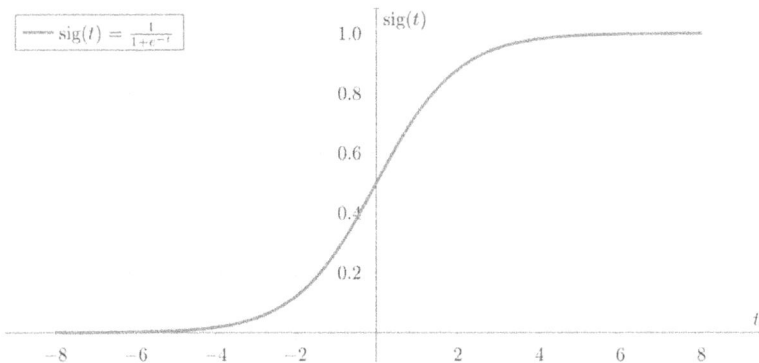

FIGURE 9.1 Sigmoid Function.

Logistic Regression Equation

The odd is the ratio of something occurring to something not occurring. It is different from probability as the probability is the ratio of something occurring to everything that could possibly occur. So odd will be:

$$\frac{p(x)}{1-p(x)} = e^z$$

Applying natural log on odd, we get:

$$\log\left[\frac{p(x)}{1-p(x)}\right] = z$$

$$\log\left[\frac{p(x)}{1-p(x)}\right] = w.X + b$$

Then the final logistic regression equation will be:

$$p(X;b,w) = \frac{e^{w.X+b}}{1+e^{w.X+b}} = \frac{1}{1+e^{-w.X+b}}$$

Likelihood Function for Logistic Regression

The predicted probabilities will be $p(X; b, w) = p(x)$ for $y = 1$ and for $y = 0$ predicted probabilities will be $1-p(X;b, w) = 1-p(x)$:

$$L(b,w) = \prod_{i=1} np(x_i)^{y_i}\left(1-p(x_i)\right)^{1-y_i}$$

Taking natural logs on both sides:

$$l(b,w) = \log\left(L(b,w)\right) = \sum_{i=1}^{n} y_i \log p(x_i) + \left(1-y_i\right)\log\left(1-p(x_i)\right)$$

$$= \sum_{i=1}^{n} y_i \log p(x_i) + \log\left(1-p(x_i)\right) - y_i \log\left(1-p(x_i)\right)$$

$$= \sum_{i=1}^{n} y_i \log\left(1-p(x_i)\right) + \sum_{i=1}^{n} y_i \log\frac{p(x_i)}{1-p(x_i)}$$

$$= \sum_{i=1}^{n} -\log 1 - e^{-(w.x_i+b)} + \sum_{i=1}^{n} y_i\left(w.x_i + b\right)$$

$$= \sum_{i=1}^{n} -\log 1 + e^{(w.x_i+b)} + \sum_{i=1}^{n} y_i\left(w.x_i + b\right)$$

Gradient of the Log-Likelihood Function

To find the maximum likelihood estimates, we differentiate with respect to w:

$$\frac{\partial J\left(l\left(b,w\right)\right)}{\partial w_j} = -\sum_{i=n}^{n}\frac{1}{1+e^{\left(w.x_i+b\right)}}e^{\left(w.x_i+b\right)}x_{ij} + \sum_{i=1}^{n}y_ix_{ij} = -\sum_{i=1}^{n}p\left(x_i;b,w\right)x_{ij}$$

$$+\sum_{i=1}^{n}y_ix_{ij} = \sum_{i=1}^{n}\left(y_i - p\left(x_i;b,w\right)\right)x_{ij}$$

9.3   RIDGE REGRESSION

Ridge regression, or Tikhonov regularization [3], is an extension of ordinary least squares (linear) regression with an additional $^c l_2$-penalty term (or ridge constraint) to regularize the regression coefficients. Machine learning models that leverage ridge regression identify the optimal set of regression coefficients as follows:

$$\hat{\beta} = \arg\min_{\beta\in\mathbb{R}^P}\left\|y - X\beta\right\|_2^2 + \lambda\left\|\beta\right\|_2^2$$

Where $y \in \mathbb{R}^N$ is the dependent/target variable whose value the model is trying to predict using N samples of training data, $X \in \mathbb{R}^{N\times P}$, and p features. The "shrinkage parameter" or "regularization coefficient," λ controls l_2-penalty term on the regression coefficient $\left\|\beta\right\|_2^2$. Increasing λ forces the regression coefficients in the AI model to become smaller. If $\lambda = 0$, the formulation is equivalent to ordinary least squares regression.

Ridge regression adds the l_2-penalty term to ensure that the linear regression coefficients do not explode (or become very large). It reduces variance, producing more consistent results on unseen datasets. It also helps deal with multicollinearity, which happens when the P features in X are linearly dependent.

Due to multicollinearity, we see a very large variance in the least square estimates of the model. So, to reduce this variance, a degree of bias is added to the regression estimates. Ordinary Least Square (OLS) will create a model by minimizing the value of Sum Square Error (SSE), whereas the Ridge regression will create a model by minimizing:

$$SSE + \lambda\sum_{i=1}^{n}\left(\beta_i\right)^2$$

9.4   LASSO REGRESSION

LASSO (Least Absolute Shrinkage Selector Operator) algorithm is another variation of linear regression like ridge regression. We use LASSO regression when we have a large number of predictor variables. The equation of LASSO is like ridge regression and looks like as given here:

$$SSE + \lambda\sum_{i=1}^{n}\left\lceil\beta_i\right\rceil$$

Here the objective is as follows: if $\lambda = 0$, we get same coefficients as linear regression. If λ is very large, all coefficients are shirked toward 0. The main difference between Ridge and LASSO Regression is that if ridge regression can shrink the coefficient close to **0** so that all predictor variables are retained, whereas LASSO can shrink the coefficient to exactly **0** so that LASSO can select and discard the predictor variables that have the right coefficient of **0**.

9.5   TREE-BASED MODELS

In short, tree-based models use a series of "if-then" rules to predict from decision trees. In this section, we'll specify commonly used linear models in machine learning, their advantages, and disadvantages.

9.5.1   DECISION TREE

Decision trees (DTs) are a non-parametric supervised learning method used for classification and regression (see Figure 9.2). The goal is to create a model that predicts the value of a target variable by learning simple decision rules inferred from the data features. A tree can be seen as a piecewise constant approximation. For instance, in the following example, decision trees learn from data to approximate a sine curve with a set of if-then-else decision rules. The deeper the tree, the more complex the decision rules and the fitter the model.
Some advantages of decision trees are:

- Simple to understand and to interpret. Trees can be visualized.
- Requires little data preparation. Other techniques often require data normalization, dummy variables need to be created and blank values to be removed. Some tree and algorithm combinations support missing values.

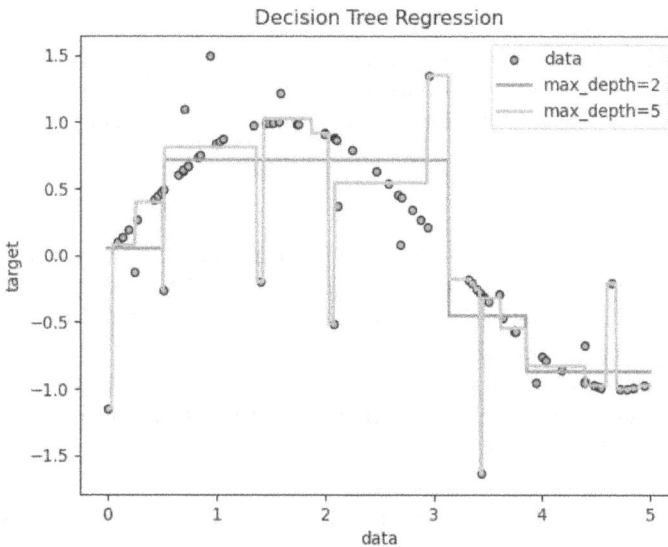

FIGURE 9.2   Decision Trees.

- The cost of using the tree (i.e., predicting data) is logarithmic in the number of data points used to train the tree.
- Able to handle both numerical and categorical data. However, the scikit-learn implementation does not support categorical variables for now. Other techniques are usually specialized in analyzing datasets that have only one type of variable.
- Able to handle multi-output problems.

If a given situation is observable in a model, the condition is easily explained by Boolean logic. By contrast, in a black box model (e.g., in an artificial neural network), results may be more difficult to interpret. Possibly, statistical tests can be used to validate a model. That makes it possible to account for the reliability of the model. The model performs well even if its assumptions are somewhat violated by the true model from which the data were generated.

The disadvantages of decision trees include:

- Decision-tree learners can create over-complex trees that do not generalize the data well. This is called overfitting. Mechanisms such as pruning, setting the minimum number of samples required at a leaf node, or setting the maximum depth of the tree are necessary to avoid this problem.
- Decision trees can be unstable because small variations in the data might result in a completely different tree being generated. This problem is mitigated by using decision trees within an ensemble.
- Predictions of decision trees are neither smooth nor continuous, but piecewise constant approximations as seen in Figure 9.2. Therefore, they are not good at extrapolation.
- The problem of learning an optimal decision tree is known to be NP-complete under several aspects of optimality and even for simple concepts. Consequently, practical decision-tree learning algorithms are based on heuristic algorithms such as the greedy algorithm where locally optimal decisions are made at each node. Such algorithms cannot guarantee to return the globally optimal decision tree. This can be mitigated by training multiple trees in an ensemble learner, where the features and samples are randomly sampled with replacement.
- There are concepts that are hard to learn because decision trees do not express them easily, such as XOR, parity, or multiplexer problems.
- Decision tree learners create biased trees if some classes dominate. It is therefore recommended to balance the dataset prior to fitting with the decision tree.

9.5.2 RANDOM FORESTS

Random forest regression is a supervised learning algorithm and bagging technique that uses an ensemble learning method for regression in machine learning. The trees in random forests run in parallel, meaning there is no interaction between these trees while building the trees. Random forest is a commonly used machine learning algorithm, which combines the output of multiple decision trees to reach a single result. Its ease of use and flexibility have fueled its adoption, as it handles both classification and regression problems. However, despite these advantages, a random forest algorithm also has some drawbacks.

- It requires much computational power as well as resources as it builds numerous trees to combine their outputs.
- Also, it requires much time for training as it combines a lot of decision trees to determine the class.

9.5.3 Gradient Boosting Regression

Gradient boosting regression trees are based on the idea of an ensemble method derived from a decision tree. The decision tree uses a tree structure. Starting from tree root, branching according to the conditions, and heading toward the leaves, the goal leaf is the prediction result. Gradient boosting is a machine learning technique used in regression and classification tasks, among others. It gives a prediction model in the form of an ensemble of weak prediction models, i.e., models that make very few assumptions about the data, which are typically simple decision trees. When a decision tree is the weak learner, the resulting algorithm is called gradient-boosted trees. It usually outperforms random forest. A gradient-boosted trees model is built in a stage-wise fashion as in other boosting methods, but it generalizes the other methods by allowing optimization of an arbitrary differentiable loss function. The algorithm suffers from a few disadvantages as follows:

- Training the model may require more time and resources compared to simpler algorithms.
- In the case of complex models or high learning rates, gradient descent can be prone to the overfitting problem.

9.5.4 XGBoost

XGBoost (eXtreme Gradient Boosting) minimizes a regularized (L1 and L2) objective function that combines a convex loss function (based on the difference between the predicted and target outputs) and a penalty term for model complexity (in other words, the regression tree functions). XGBoost is designed for efficient and scalable training of machine learning models, making it suitable for large datasets. XGBoost has a wide range of hyperparameters that can be adjusted to optimize performance, making it highly customizable. Tree algorithms such as XGBoost and Random Forest do not need normalized features and work well if the data is nonlinear, non-monotonic, or with segregated clusters. However, Tree algorithms such as XGBoost and Random Forest can overfit the data, especially if the trees are too deep with noisy data.

9.5.5 Light Gradient Boosted Machine Regressor

Light Gradient Boosted Machine (LightGBM) is a gradient boosting ensemble method that is an open-source library that provides an efficient and effective implementation of the gradient boosting algorithm. It is based on decision trees. As with other decision tree-based methods, LightGBM can be used for both classification and regression. LightGBM is optimized for high performance with distributed systems. LightGBM extends the gradient boosting algorithm by adding a type of automatic feature selection as well as focusing on boosting examples with larger gradients. This can result in a dramatic speedup of training and improved predictive performance. As such, LightGBM has become a de facto algorithm for machine learning competitions when working with tabular data for regression and classification predictive modeling tasks. Thus, it owns a share of the blame for the increased popularity and wider adoption of gradient boosting methods in general, along with Extreme Gradient Boosting (XGBoost).

9.6 SUMMARY

We have described the characteristics of supervised machine learning, which uses labeled data, as well as linear regression and classification, logistic regression, ridge regression, LASSO regression, and tree-based algorithms that are popularly used for supervised machine learning. The decision tree, random forests, gradient boosting regression, XGBoost, and LightGBM regressor that are part of the tree-based models are also explained. These algorithms have open libraries that are widely over the Internet and are extensively used by users worldwide.

REFERENCES

[1] Poole, D. L. and Mackworth, A. K., "Artificial Intelligence: Foundation of Computational Agents," Second Edition, 2017, Cambridge University Press.
[2] https://www.geeksforgeeks.org/understanding-logistic-regression/.
[3] https://c3.ai/glossary/data-science/ridge-regression/.

10 Deep Learning

Deep learning uses artificial neural networks to mimic the learning process of the human brain. Neural network is also termed as artificial neural network (ANN). A deep neural network can be characterized by a succession of multiple processing layers. Each layer consists of a nonlinear transformation with activation function (e.g., sigmoid and rectified linear unit—ReLU), and the sequence of these transformations leads to learning different levels of abstraction. In general, deep learning (DL) relies on a function $f: X \rightarrow Y$ parameterized with:

$$\theta \in R^{n_\theta} \text{ and } n_g \in N: \quad y = f(x;\theta)$$

We are describing a very simple neural network with one fully connected hidden layer (see Figure 10.1). The first layer is given the input values (i.e., the input features) x in the form of a column vector of size n_x and $n_x \in N$. The values of the next hidden layer are a transformation of these values by a nonlinear parametric function, which is a matrix multiplication by W_1 of size $n_h \times n_x$ and $n_x \in N$, plus a bias term b_2 of size n_h, followed by a nonlinear transformation: [1]

$$h = A(W_1.x + b_1)$$

where A is an activation function, and we are using a sigmoid function activation function: $f(x) = (1 + e^{-x})^{-1}$

This nonlinear activation function is what makes the transformation at each layer nonlinear, which ultimately provides the expressivity of the neural network. The hidden layer h of size n_h can in turn be transformed to other sets of values up to the last transformation that provides the output values y (as depicted in Figure 10.1). In this case:

$$y = A(W_2.h + b_2)$$

where W_2 of size $n_y \times n_h$ and $n_y \in N$, plus a bias term b_2 of size $n_y \in N$. All these layers are trained to minimize the empirical error $e_S[f]$. The most common method for optimizing the parameters of a neural network is based on gradient descent via the backpropagation algorithm. In the simplest case, at every iteration, the algorithm changes its internal parameters θ to fit the desired function:

$$\theta \leftarrow (\theta - \alpha \nabla_\theta e_S[f])$$

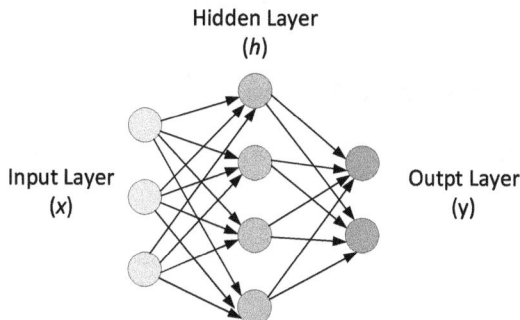

FIGURE 10.1 Example of a Fully Connected Neural Feedforward Network with One Hidden Layer.

DOI: 10.1201/9781003499466-10

Where α is the learning rate. In current applications, many different types of neural network layers have appeared beyond the simple feedforward networks just introduced. Each variation provides specific advantages, depending on the application (e.g., good tradeoff between bias and overfitting in a supervised learning setting). In addition, within one given neural network, an arbitrarily large number of layers is possible, and the trend in the last few years is to have an ever-growing number of layers, with more than 100 in some supervised learning tasks.

10.1 COST AND ERROR FUNCTION, GRADIENT DECENT, AND BACKPROPAGATION IN NEURAL NETWORK

Previously, we've talked about iteratively minimizing the error of the output of the neural network by varying the weights in gradient descent. However, as it turns out, there is a mathematically more generalized way of looking at things that allows us to reduce the error while also preventing things like overfitting (this will be discussed more in later chapter). This more general optimization formulation revolves around minimizing what's called the cost function. The equivalent cost function (also called sum-of-squares-error) of a single training pair (x^z, y^z) in a neural network is:

$$J\left(w,b,x,y\right)=\frac{1}{2}\left\|y^z-h^{(n_l)}\left(x^z\right)\right\|^2=\frac{1}{2}\left\|y^z-y_{pred}\left(x^z\right)\right\|^2$$

This shows the cost function of the z_{th} training sample, where $h^{(n_l)}$ is the output of the final layer of the neural network i.e., the output of the neural network. We have also represented $h^{(n_l)}$ as y_{pred} to highlight the prediction of the neural network given x^z. The two vertical lines represent the L_2 norm of the error or what is known as the sum-of-squares-error (SSE). SSE is a very common way of representing the error of a machine learning system. Instead of taking just the absolute error $abs\left(y^z-y_{pred}\left(x^z\right)\right)$, we use the square of the error. There are many reasons why the SSE is often used which will not be discussed here—suffice to say that this is a very common way of representing the errors in machine learning. The $\frac{1}{2}$ out the front is just a constant added that tidies things up when we differentiate the cost function, which we'll be doing when we perform backpropagation. Note that the formulation for the earlier cost function is for a single (x, y) training pair. We want to minimize the cost function over all our m training pairs. Therefore, we want to find the minimum mean squared error (MSE) over all the training samples:

$$J\left(w,b\right)=\frac{1}{m}\sum_{z=0}^{m}\frac{1}{2}\left\|y^z-h^{(n_l)}\left(x^z\right)\right\|^2\cong\frac{1}{m}\sum_{z=0}^{m}J\left(W,b,x^z,y^z\right)$$

So, how do you use the cost function J to train the weights of our network? Using gradient descent and backpropagation. First, let's look at gradient descent more closely in neural networks.

10.1.1 GRADIENT DESCENT IN NEURAL NETWORKS

Gradient descent for every weight $w_{ij}^{(l)}$ and every bias $b_i^{(l)}$ in the neural network looks like the following:

$$w_{ij}^{(l)}\leftarrow w_{ij}^{(l)}-\alpha\frac{\partial}{\partial w_{ij}^{(l)}}J\left(w,b\right)$$

$$b_i^{(l)}\leftarrow b_i^{(l)}-\alpha\frac{\partial}{\partial b_i^{(l)}}J\left(w,b\right)$$

10.1.2 TWO-DIMENSIONAL GRADIENT DESCENT EXAMPLE AND BACKPROPAGATION DEPTH

We have taken a simple example of two-dimensional gradient descent as depicted in Figure 10.2.

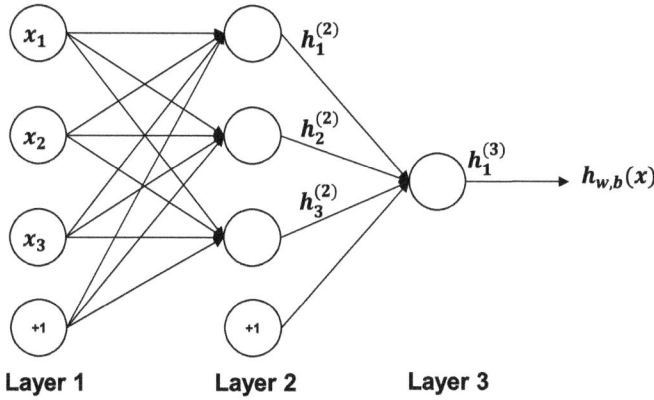

FIGURE 10.2 Simple Two-Dimensional Gradient Decent Example.

$$h_1^{(2)} = f\left(w_{11}^{(1)}x_1 + w_{12}^{(1)}x_2 + w_{13}^{(1)}x_3 + b_1^{(1)}\right) \triangleq f\left(z_1^{(2)}\right)$$

$$h_2^{(2)} = f\left(w_{21}^{(1)}x_1 + w_{22}^{(1)}x_2 + w_{23}^{(1)}x_3 + b_2^{(1)}\right) \triangleq f\left(z_2^{(2)}\right)$$

$$h_3^{(2)} = f\left(w_{31}^{(1)}x_1 + w_{32}^{(1)}x_2 + w_{33}^{(1)}x_3 + b_3^{(1)}\right) \triangleq f\left(z_3^{(2)}\right)$$

$$h_{W,b}(x) = h_1^{(3)} = f\left(w_{11}^{(2)}h_1^{(2)} + w_{12}^{(2)}h_2^{(2)} + w_{13}^{(2)}h_3^{(2)} + b_1^{(2)}\right)$$

10.1.3 VECTORIZATION IN NEURAL NETWORKS

There is a way to write the equations even more compactly and to calculate the feed forward process in neural networks more efficiently from a computational perspective. First, we can introduce a new variable $z_i^{(l)}$, which is the summated input into node i of layer l, including the bias term. So in the case of the first node in layer 2, $z_1^{(2)}$ is equal to:

$$z_i^{(2)} \triangleq \left(w_{i1}^{(1)}x_1 + w_{i2}^{(1)}x_2 + w_{i3}^{(1)}x_3 + \dots w_{in}^{(1)}x_{in} + b_i^{(1)}\right) = \sum_{j=1}^{n} w_{ij}^{(1)}x_i + b_i^{(1)}$$

where n is the number of nodes in layer 1.

Using this notation, the unwieldy previous set of equations for the example three-layer network can be reduced to:

$$z^{(2)} = W^{(1)}x + b^{(1)}$$

$$h^{(2)} = f\left(z^{(2)}\right)$$

$$z^{(3)} = W^{(2)}h^{(2)} + b^{(2)}$$

$$h_{W,b}(x) = h^{(3)} = f\left(z^{(3)}\right)$$

Note the use of capital W to denote the matrix form of the weights. It should be noted that all of the elements in the previous equation are now matrices/vectors. If you're unfamiliar with these concepts, they will be explained more fully in the next section. Can the above equation be simplified even further? Yes, it is possible. We can forward propagate the calculations through any number of layers in the neural network by generalizing (the input layer: $h^{(1)} = x$):

$$z^{(l+1)} = W^{(l)}h^{(l)} + b^{(l)}$$

$$h^{(l+1)} = f\left(z^{(l+1)}\right)$$

Here we can see the general feed forward process, where the output of layer l becomes the input to layer $l + 1$. We know that $h^{(1)}$ is simply the input layer x, and $h^{(n_l)}$ (where n_l is the number of layers in the network) is the output of the output layer. Notice that in the previous equations we have dropped references to the node numbers i and j—how can we do this? Don't we still have to loop through and calculate all the various node inputs and outputs?

The answer is that we can use matrix multiplications to do this more simply. This process is called "vectorization," and it has two benefits—first, it makes the code less complicated, as you will see shortly. Second, we can use fast linear algebra routines in Python (and other languages) rather than using loops, which will speed up our programs. Numpy can handle these calculations easily. First, for those who aren't familiar with matrix operations, the next section is a brief recap.

10.1.4 MATRIX MULTIPLICATION

$$z^{(l+1)} = W^{(l)}x + b^{(l)}$$

However, $l = 1$, i.e., input layer $h^{(l)} = h^{(1)} \triangleq x$

$$z^{(2)} = \begin{bmatrix} w_{11}^{(1)} & w_{12}^{(1)} & w_{13}^{(1)} \\ w_{21}^{(1)} & w_{22}^{(1)} & w_{23}^{(1)} \\ w_{31}^{(1)} & w_{32}^{(1)} & w_{33}^{(1)} \end{bmatrix} \begin{bmatrix} x_1 \\ x_2 \\ x_3 \end{bmatrix} + \begin{bmatrix} b_1^{(1)} \\ b_2^{(1)} \\ b_3^{(1)} \end{bmatrix}$$

$$W^{(1)} = \begin{bmatrix} w_{11}^{(1)} & w_{12}^{(1)} & w_{13}^{(1)} \\ w_{21}^{(1)} & w_{22}^{(1)} & w_{23}^{(1)} \\ w_{31}^{(1)} & w_{32}^{(1)} & w_{33}^{(1)} \end{bmatrix}$$

$$b^{(1)} = \begin{bmatrix} b_1^{(1)} \\ b_2^{(1)} \\ b_3^{(1)} \end{bmatrix}, \quad x = \begin{bmatrix} x_1 \\ x_2 \\ x_3 \end{bmatrix}$$

Only for input layer:

$$h^{(2)} = W^{(1)} * x + b^{(1)}$$

If like we can write for layer 2 weight in a similar way as follows:

$$W^{(2)} = \begin{bmatrix} w_{11}^{(2)} & w_{12}^{(2)} & w_{13}^{(2)} \end{bmatrix}$$

Using this notation, the unwieldy previous set of equations, for example, a three-layer network, can be reduced to:

$$z^{(2)} = W^{(1)}x + b^{(1)}$$

$$h^{(3)} = f\left(z^{(2)}\right)$$

$$z^{(3)} = W^{(2)}h^{(2)} + b^{(2)}$$

$$h_{W,b}(x) = h^{(3)} = f\left(z^{(3)}\right)$$

$$l = 1 \rightarrow \frac{\partial}{\partial w_{ij}^{(1)}} J\left(w,b,x,y\right) = \delta_i^{(2)}\left(h^{(1)}\right)^T$$

$$l = 2 \rightarrow \frac{\partial}{\partial w_{ij}^{(2)}} J\left(w,b,x,y\right) = \delta_i^{(3)}\left(h^{(2)}\right)^T$$

$$l = 3 \rightarrow \frac{\partial}{\partial w_{ij}^{(3)}} J\left(w,b,x,y\right) = \delta_i^{(4)}\left(h^{(3)}\right)^T$$

$$\frac{\partial}{\partial w_{ij}^{(l)}} J\left(w,b,x,y\right) = \delta_i^{(l+1)}\left(h^{(l)}\right)^T$$

We have found for the term: $h_{w,b}(x) = h^{(3)} = f(z^{(2)})$. Let us generalize our equations for this given neural network configuration:

Only for input layer:

$$z^{(l+1)} = W^{(l)}x + b^{(l)}$$

For all other layers (other than input layer) for this configuration:

$$h^{(l+1)} = f(z^{(l+1)})$$

10.1.5   Cost/Error/Loss Function Derivatives for Gradient Descent

Let us evaluate this specific example: $\frac{\partial}{\partial w_{12}^{(2)}} J(w,b)$. To do so, we must develop a chain function:

$$\frac{\partial}{\partial w_{12}^{(2)}} J(w,b) = \frac{\partial J}{\partial h_1^{(3)}} \frac{\partial h_1^{(3)}}{\partial z_1^{(3)}} \frac{\partial z_1^{(3)}}{\partial w_{12}^{(2)}}$$

$$\frac{\partial z_1^{(3)}}{\partial w_{12}^{(2)}} = \frac{\partial}{\partial w_{12}^{(2)}}\left(w_{11}^{(2)}h_1^{(2)} + w_{12}^{(2)}h_2^{(2)} + w_{13}^{(2)}h_3^{(2)} + b_1^{(2)} \right)$$

$$= \frac{\partial}{\partial w_{12}^{(2)}}\left(w_{12}^{(2)}h_2^{(2)} \right) = h_2^{(2)}$$

$h_2^{(2)}$ is simply the output of the second node in layer 2. Next term is this: $\frac{\partial h_1^{(3)}}{\partial z_1^{(2)}}$, where $h_1^{(3)}(z)$ is the sigmoid activation function, and the differentiation of this activation function is as follows (see earlier derivation):

$$\frac{\partial h(z)}{\partial z} = f'(z) = f(z)\left(1 - f(z)\right)$$

$$\frac{\partial h_1^{(3)}}{\partial z_1^{(3)}} = \frac{\partial}{\partial z_1^{(3)}}\left[f\left(w_{11}^{(2)}h_1^{(2)} + w_{12}^{(2)}h_2^{(2)} + w_{13}^{(2)}h_3^{(2)} + b_1^{(2)} \right) \right] = \frac{\partial}{\partial z_1^{(3)}}\left[f\left(z_1^{(3)} \right) \right] = f'\left(z_1^{(3)} \right)$$

$$= f\left(z_1^{(3)} \right)\left(1 - f\left(z_1^{(3)} \right) \right)$$

The final term is $\frac{\partial J}{\partial h_1^{(3)}}$ where $J(w,b,x,y)$ is the mean-square-error loss function, which looks like (for our case):

$$J(w,b) = \frac{1}{m}\sum_{z=0}^{m}\frac{1}{2}\left\| y^z - h^{(n_l)}\left(x^z \right) \right\|^2 \triangleq \frac{1}{m}\sum_{z=0}^{m} J\left(W,b,x^z,y^z \right)$$

$$J(w,b,x,y) = \frac{1}{2}\left\| y_1 - \hat{y}_1 \right\|^2 = \frac{1}{2}\left\| y_1 - h_1^{(3)} \right\|^2$$

where,

y_1 is the target output.

$\hat{y}_1$ is the predicted output $= h_1^{(3)} \triangleq f\left(z_1^{(3)} \right) \triangleq h_1^{(3)}\left(z_1^{(3)} \right)$

Let

$$u \triangleq \left\| y_1 - h_1^{(3)} \right\|$$

So,

$$J(w,b,x,y) = \frac{1}{2}u^2$$

$$\frac{\partial J}{\partial h_1^{(3)}} = \left[\frac{\partial J}{\partial u}\right]\left[\frac{\partial u}{\partial h_1^{(3)}}\right] = \left[\frac{\partial u}{\partial h_1^{(3)}}\right]\left[\frac{\partial J}{\partial u}\right] = \left[\frac{\partial u}{\partial h_1^{(3)}}\right]\left[\frac{\partial}{\partial u}\left(\frac{1}{2}u^2\right)\right]$$

$$= u\left[\frac{\partial u}{\partial h_1^{(3)}}\right] = u\left[\frac{\partial}{\partial h_1^{(3)}}\left(\left\| y_1 - h_1^{(3)} \right\| \right)\right] = -u = -\left\| y_1 - h_1^{(3)} \right\|$$

$$\frac{\partial J}{\partial h_1^{(3)}} = -\left\| y_1 - h_1^{(3)} \right\|$$

Normalizing L2 norm: Normalizing a vector: $\dfrac{x}{\|x\|}$ where $\|x\| = \left(x_1^2 + x_2^2 + \ldots + x_n^2 \right)^{1/2}$

$$\|x\| = \left(x_1^2 + x_2^2 + \ldots + x_n^2 \right)^{1/2}$$

If $\|x\| = 1$, it is called unity norm.

$$\frac{\partial J}{\partial h_1^{(3)}} = -\left\| y_1 - h_1^{(3)} \right\|$$

We can generalize this equation:

$$\frac{\partial J}{\partial h_i^{(n_l)}} = -\left(y_i - h_i^{(n_l)} \right)$$

$$\frac{\partial}{\partial w_{12}^{(2)}} J(w, b, x, y) = \left[\frac{\partial J}{\partial h_1^{(3)}} \right] \left[\frac{\partial h_1^{(3)}}{\partial z_1^{(3)}} \right] \left[\frac{\partial z_1^{(3)}}{\partial w_{12}^{(2)}} \right]$$

$$= -\left[y_1 - h_1^{(3)} \right] \left[f'\left(z_1^{(3)} \right) \right] \left[\frac{\partial z_1^{(3)}}{\partial w_{12}^{(2)}} \right] = \delta_1^{(3)} * \left[\frac{\partial z_1^{(3)}}{\partial w_{12}^{(2)}} \right] = \delta_1^{(3)} * h_2^{(2)}$$

where $\delta_1^{(3)} \triangleq -\left(y_1 - h_1^{(3)} \right) f'\left(z_1^{(3)} \right)$

Note: $\dfrac{\partial h_1^{(3)}}{\partial z_1^{(3)}} = \dfrac{\partial f\left(z_1^{(3)} \right)}{\partial z_1^{(3)}} = f'\left(z_1^{(3)} \right)$

Generalizing this equation, we can write:

$$\delta_i^{(n_l)} \triangleq -\left(y_i - h_i^{(n_l)} \right) f'\left(z_i^{(n_l)} \right) = -\left(y_i - f\left(z_i^{(n_l)} \right) \right) f'\left(z_i^{(n_l)} \right)$$

where i is the ith node of the output layer. For example, in the given configuration that we have shown here, it has always been $i = 1$.

$$\frac{\partial}{\partial w_{12}^{(2)}} J(w, b, x, y) = \delta_1^{(3)} * h_2^{(2)}$$

Generalizing this equation, we have:

$$\frac{\partial}{\partial w_{ij}^{(l)}} J(w, b, x, y) = \delta_i^{(l+1)} * h_j^{(l)}$$

n_l = Number of nodes in layer l.

i = Number of nodes in layer $l + 1$. In this neural network configuration, there is only one node in layer 3. So, $i = 1$ only.

j = Number of nodes in layer l.

10.1.6 PROPAGATING INTO HIDDEN LAYERS

What about for weights feeding into any hidden layers (layer 2 in our case)? For the weights connecting the output layer, the following derivative made sense, as the cost function can be directly calculated by comparing the output layer to the training data:

$$\frac{\partial J}{\partial h_i^{(n_l)}} = -\left(y_i - h_i^{(n_l)}\right)$$

The output of the hidden nodes, however, has no such direct reference, rather they are connected to the cost function only through mediating weights and potentially other layers of nodes. How can we find the variation in the cost function from changes to weights embedded deep within the neural network? As mentioned previously, we use the backpropagation method. Now that we've done the demanding work using the chain rule, we'll now take a more graphical approach. The term that needs to propagate back through the network is the $\delta_i^{(n_l)}$ term, as this is the network's ultimate connection to the cost function.

What about node j in the second layer (hidden layer)? How does it contribute to $\delta_i^{(n_l)}$ in our test network? It contributes via the weight $w_{ij}^{(2)}$—see Figure 10.3 for the case of $j = 1$ and $i = 1$.

As can be observed from the figure, the output layer δ is communicated to the hidden node by the weight of the connection. In the case where there is only one output layer node, the generalized hidden layer δ is defined as:

$$\delta_j^{(l)} = \left(w_{1j}^{(l)}\delta_1^{(l+1)}\right)f'\left(z_j\right)^{(l)}$$

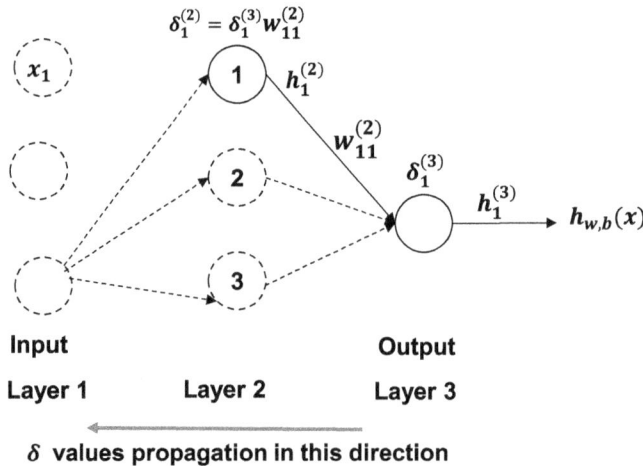

FIGURE 10.3 Back Propagation Model.

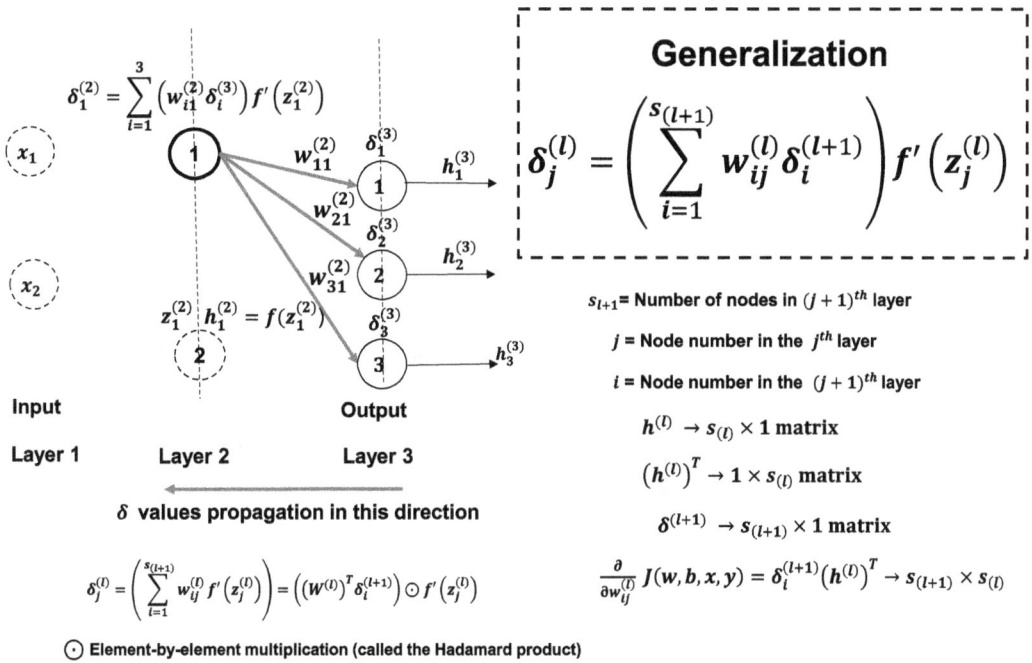

FIGURE 10.4 Generalization of back propagation model.

where j is the node number in layer l.

What about the case where there are multiple output nodes? In this case, the weighted sum of all the communicated errors is taken to calculate $\delta_j^{(l)}$, as shown in Figure 10.4.

As can be observed from the figure, each δ value from the output layer is included in the sum used to calculate $\delta_1^{(2)}$, but each output δ is weighted according to the appropriate $w_{i1}^{(2)}$ value. In other words, node 1 in layer 2 contributes to the error of three output nodes, therefore the measured error (or cost function value) at each of these nodes must be "passed back" to the δ value for this node. Now we can develop a generalized expression for the δ values for nodes in the hidden layers:

$$\delta_j^{(l)} = \left(\sum_{i=1}^{s_{(l+1)}} w_{ij}^{(l)} \delta_i^{(l+1)} \right) f'\left(z_j^{(l)}\right) = \left(\left(W^{(l)}\right)^T \delta_i^{(l+1)} \right) \odot f'\left(z_j^{(l)}\right)$$

where j is the node number in layer l, and i is the node number in layer $l + 1$ (which is the same notation, we have used from the start). The value $s_{(l+1)}$ is the number of nodes in layer $(l + 1)$. Note: $\odot$ is element-by-element multiplication called Hadamard product. Finally, a complete picture of a similar neural network with four layers can be described in Figure 10.5.

Layers 2 and 3 are called hidden layers.

$$z^{(2)} = \begin{bmatrix} w_{11}^{(1)} & w_{12}^{(1)} & w_{13}^{(1)} \\ w_{21}^{(1)} & w_{22}^{(1)} & w_{23}^{(1)} \\ w_{31}^{(1)} & w_{32}^{(1)} & w_{33}^{(1)} \end{bmatrix} \begin{bmatrix} x_1 \\ x_2 \\ x_3 \end{bmatrix} + \begin{bmatrix} b_1^{(1)} \\ b_2^{(1)} \\ b_3^{(1)} \end{bmatrix}$$

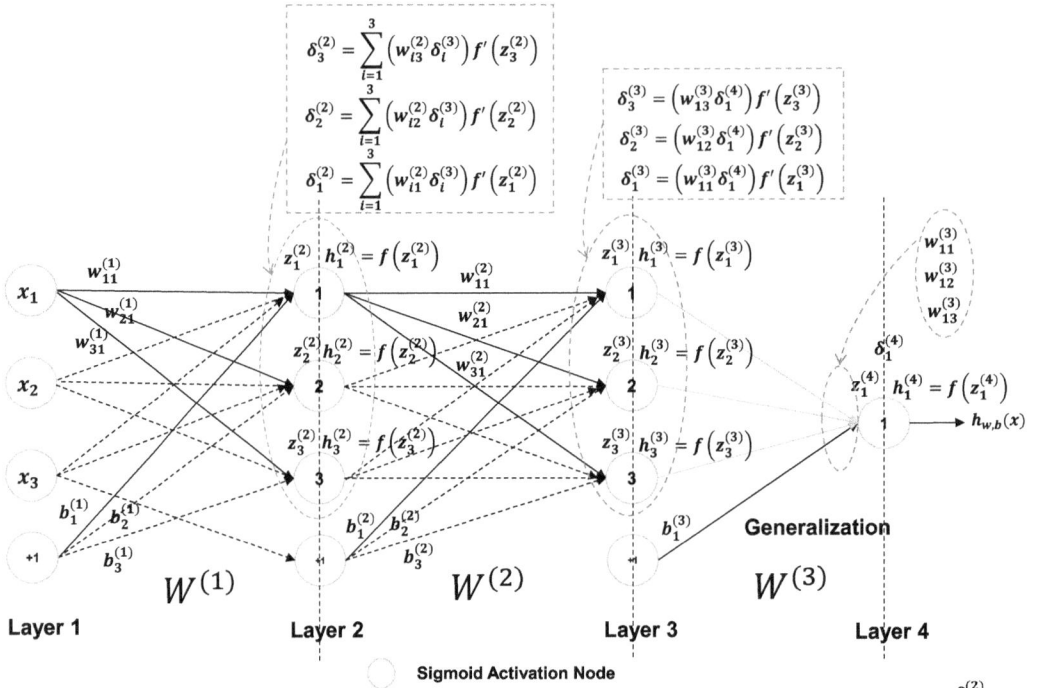

$$\delta_3^{(2)} = \sum_{i=1}^{3}\left(w_{i3}^{(2)}\delta_i^{(3)}\right)f'\left(z_3^{(2)}\right)$$
$$\delta_2^{(2)} = \sum_{i=1}^{3}\left(w_{i2}^{(2)}\delta_i^{(3)}\right)f'\left(z_2^{(2)}\right)$$
$$\delta_1^{(2)} = \sum_{i=1}^{3}\left(w_{i1}^{(2)}\delta_i^{(3)}\right)f'\left(z_1^{(2)}\right)$$

$$\delta_3^{(3)} = \left(w_{13}^{(3)}\delta_1^{(4)}\right)f'\left(z_3^{(3)}\right)$$
$$\delta_2^{(3)} = \left(w_{12}^{(3)}\delta_1^{(4)}\right)f'\left(z_2^{(3)}\right)$$
$$\delta_1^{(3)} = \left(w_{11}^{(3)}\delta_1^{(4)}\right)f'\left(z_1^{(3)}\right)$$

Sigmoid Activation Node

$$\delta_3^{(2)}$$

s_{l+1} = Number of nodes in $(j+1)^{th}$ layer

j = Node number in the jth layer

i = Node number in the $(j+1)^{th}$ layer

$h^{(l)} \rightarrow s_{(l)} \times 1$ matrix

$\left(h^{(l)}\right)^T \rightarrow 1 \times s_{(l)}$ matrix

$\delta^{(l+1)} \rightarrow s_{(l+1)} \times 1$ matrix

$$\frac{\partial}{\partial w_{ij}^{(l)}} J(w,b,x,y) = \delta_i^{(l+1)}\left(h^{(l)}\right)^T \rightarrow s_{(l+1)} \times s_{(l)}$$

$$\delta_j^{(l)} = \left(\sum_{i=1}^{s_{(l+1)}} w_{ij}^{(l)} f'\left(z_j^{(l)}\right)\right) = \left(\left(W^{(l)}\right)^T \delta_i^{(l+1)}\right) \odot f'\left(z_j^{(l)}\right)$$

$\odot$ **Element-by-element multiplication (called the Hadammard product)**

FIGURE 10.5   Back Propagation Model in Each Neural Network Layer.

$$W^{(1)} = \begin{bmatrix} w_{11}^{(1)} & w_{12}^{(1)} & w_{13}^{(1)} \\ w_{21}^{(1)} & w_{22}^{(1)} & w_{23}^{(1)} \\ w_{31}^{(1)} & w_{32}^{(1)} & w_{33}^{(1)} \end{bmatrix}$$

$$b^{(1)} = \begin{bmatrix} b_1^{(1)} \\ b_2^{(1)} \\ b_3^{(1)} \end{bmatrix}, \qquad x = \begin{bmatrix} x_1 \\ x_2 \\ x_3 \end{bmatrix}$$

$$\delta^{(1)} = \begin{bmatrix} \delta_1^{(1)} \\ \delta_2^{(1)} \\ \delta_3^{(1)} \end{bmatrix} = \begin{bmatrix} w_{11}^{(1)} & w_{12}^{(1)} & w_{13}^{(1)} \\ w_{21}^{(1)} & w_{22}^{(1)} & w_{23}^{(1)} \\ w_{31}^{(1)} & w_{32}^{(1)} & w_{33}^{(1)} \end{bmatrix}^T \begin{bmatrix} \delta_1^{(2)} \\ \delta_2^{(2)} \\ \delta_3^{(2)} \end{bmatrix} \begin{bmatrix} f'\left(z_1^{(1)}\right) \\ f'\left(z_2^{(1)}\right) \\ f'\left(z_3^{(1)}\right) \end{bmatrix}$$

Θ = Element-by-Element Multiplication Hadamard Matrix

$$\delta_1^{(1)} = w_{11}^{(1)}\delta_1^{(2)}f'\left(z_1^{(1)}\right) + w_{21}^{(1)}\delta_2^{(2)}f'\left(z_1^{(1)}\right) + w_{31}^{(1)}\delta_3^{(2)}f'\left(z_1^{(1)}\right)$$

$$\delta_2^{(1)} = w_{12}^{(1)}\delta_1^{(2)}f'\left(z_2^{(1)}\right) + w_{22}^{(1)}\delta_2^{(2)}f'\left(z_2^{(1)}\right) + w_{32}^{(1)}\delta_3^{(2)}f'\left(z_2^{(1)}\right)$$

$$\delta_3^{(1)} = w_{13}^{(1)}\delta_1^{(2)}f'\left(z_3^{(1)}\right) + w_{23}^{(1)}\delta_2^{(2)}f'\left(z_{33}^{(1)}\right) + w_{31}^{(1)}\delta_3^{(2)}f'\left(z_3^{(1)}\right)$$

$$z^{(3)} = \begin{bmatrix} w_{11}^{(2)} & w_{12}^{(2)} & w_{13}^{(2)} \\ w_{21}^{(2)} & w_{22}^{(2)} & w_{23}^{(2)} \\ w_{31}^{(2)} & w_{32}^{(2)} & w_{33}^{(2)} \end{bmatrix}\begin{bmatrix} h_1^{(2)} \\ h_2^{(2)} \\ h_3^{(2)} \end{bmatrix} + \begin{bmatrix} b_1^{(2)} \\ b_2^{(2)} \\ b_3^{(2)} \end{bmatrix}$$

$$W^{(2)} = \begin{bmatrix} w_{11}^{(2)} & w_{12}^{(2)} & w_{13}^{(2)} \\ w_{21}^{(2)} & w_{22}^{(2)} & w_{23}^{(2)} \\ w_{31}^{(2)} & w_{32}^{(2)} & w_{33}^{(2)} \end{bmatrix}$$

$$\delta^{(2)} = \begin{bmatrix} \delta_1^{(2)} \\ \delta_2^{(2)} \\ \delta_3^{(2)} \end{bmatrix} = \begin{bmatrix} \begin{bmatrix} w_{11}^{(2)} & w_{12}^{(2)} & w_{13}^{(2)} \\ w_{21}^{(2)} & w_{22}^{(2)} & w_{23}^{(2)} \\ w_{31}^{(2)} & w_{32}^{(2)} & w_{33}^{(2)} \end{bmatrix}^{T}\begin{bmatrix} \delta_1^{(3)} \\ \delta_2^{(3)} \\ \delta_3^{(3)} \end{bmatrix} \end{bmatrix} \begin{bmatrix} f'\left(z_1^{(2)}\right) \\ f'\left(z_2^{(2)}\right) \\ f'\left(z_3^{(2)}\right) \end{bmatrix}$$

Θ = Element-by-Element Multiplication Hadamard Matrix

$$\delta_1^{(2)} = w_{11}^{(2)}\delta_1^{(3)}f'\left(z_1^{(2)}\right) + w_{21}^{(2)}\delta_2^{(3)}f'\left(z_1^{(2)}\right) + w_{31}^{(2)}\delta_3^{(3)}f'\left(z_1^{(2)}\right)$$

$$\delta_2^{(2)} = w_{12}^{(2)}\delta_1^{(3)}f'\left(z_2^{(2)}\right) + w_{22}^{(2)}\delta_2^{(3)}f'\left(z_2^{(2)}\right) + w_{32}^{(2)}\delta_3^{(3)}f'\left(z_2^{(2)}\right)$$

$$\delta_3^{(2)} = w_{13}^{(2)}\delta_1^{(3)}f'\left(z_3^{(2)}\right) + w_{23}^{(2)}\delta_2^{(3)}f'\left(z_{33}^{(2)}\right) + w_{31}^{(2)}\delta_3^{(3)}f'\left(z_3^{(2)}\right)$$

$$b^{(2)} = \begin{bmatrix} b_1^{(2)} \\ b_2^{(2)} \\ b_3^{(2)} \end{bmatrix}, \quad h^{(2)} = \begin{bmatrix} h_1^{(2)} \\ h_2^{(2)} \\ h_3^{(2)} \end{bmatrix}$$

$$z^{(4)} = \begin{bmatrix} w_{11}^{(3)} & w_{12}^{(3)} & w_{13}^{(3)} \end{bmatrix}\begin{bmatrix} h_1^{(3)} \\ h_2^{(3)} \\ h_3^{(3)} \end{bmatrix} + \begin{bmatrix} b_1^{(3)} \end{bmatrix}$$

$$W^{(3)} = \begin{bmatrix} w_{11}^{(3)} & w_{12}^{(3)} & w_{13}^{(3)} \end{bmatrix}$$

$$b^{(3)} = \begin{bmatrix} b_1^{(3)} \end{bmatrix}, \qquad h^{(3)} = \begin{bmatrix} h_1^{(3)} \\ h_2^{(3)} \\ h_3^{(3)} \end{bmatrix}$$

$$\delta^{(3)} = \begin{bmatrix} w_{11}^{(3)} * f'\left(z_1^{(3)}\right) & w_{12}^{(3)} * f'\left(z_2^{(3)}\right) * & w_{13}^{(3)} * f'\left(z_3^{(3)}\right) \end{bmatrix}^T * \delta_1^{(4)}$$

Since we have:

$$\delta_i^{(n_l)} \triangleq -\left(y_i - h_i^{(n_l)}\right) f'\left(z_i^{(n_l)}\right)$$

So, for 4-node neural output layer:

$$\delta_1^{(4)} \triangleq -\left(y_1 - h_1^{(4)}\right) f'\left(z_1^{(4)}\right) = -\left(y_1 - h_1^{(\mathrm{out})}\right) f'\left(z_1^{(\mathrm{out})}\right)$$

$$W^{(1)} = 3\times 3, \quad W^{(2)} = 3\times 3, \quad W^{(3)} = 1\times 3$$

$$h^{(1)} = x = 3\times 1, \quad h^{(2)} = 3\times 1, \quad h^{(3)} = 3\times 1 \quad h^{(4)} = 1\times 1$$

$$b^{(1)} = 3\times 1, \quad b^{(2)} = 3\times 1, \quad b^{(3)} = 1\times 1$$

$$z^{(2)} = 3\times 3, \quad z^{(3)} = 3\times 3, \quad z^{(4)} = 1\times 3$$

$$\delta^{(2)} = 3\times 1, \quad \delta^{(3)} = 3\times 1, \quad \delta^{(4)} = 1\times 1$$

$$\frac{\partial}{\partial w_{ij}^{(1)}} J\left(w,b,x,y\right) = \delta_i^{(2)}\left(h^{(1)}\right)^T$$

$$\frac{\partial}{\partial w_{ij}^{(2)}} J\left(w,b,x,y\right) = \delta_i^{(3)}\left(h^{(2)}\right)^T$$

$$\frac{\partial}{\partial w_{ij}^{(3)}} J\left(w,b,x,y\right) = \delta_i^{(4)}\left(h^{(3)}\right)^T$$

Generalizing:

$$h^{(l)} \rightarrow s_l \times 1 \ \text{matrix}$$

$$\delta^{(l+1)} \rightarrow s_{l+1} \times 1 \ \text{matrix}$$

where, S_{l+1} is the number of nodes in layer $l + 1$

However,

$$\left\{ \frac{\partial}{\partial w_{ij}^{(l)}} J\left(w,b,x,y\right) = \delta_i^{(l+1)} h^{(l)} \right\}$$

In the matrix form, $\delta_i^{(l+1)} h^{(l)}$ cannot be multiplied because the dimension does NOT match between $(s_{l+1} \times 1)$ and $(s_l \times 1)$ for multiplication unless transport matrix is formed. So we can do the following:

$$\delta_i^{(l+1)} \left(h^{(l)} \right)^T \rightarrow \left(s_{l+1} \times 1 \right) \times \left(1 \times s_l \right) \rightarrow \left(s_{l+1} \times s_l \right)$$

Finally, in the matrix form, we have:

$$\frac{\partial}{\partial w_{ij}^{(l)}} J\left(w,b,x,y\right) = \delta_i^{(l+1)} \left(h^{(l)} \right)^T$$

So we now know how to calculate:

$$\frac{\partial}{\partial w_{ij}^{(l)}} J\left(w,b,x,y\right) = \delta_i^{(l+1)} \left(h^{(l)} \right)^T \rightarrow \left(s_{l+1} \times s_l \right)$$

as shown previously. What about the bias weights? We are not going to derive them as we did with the normal weights in the interest of saving time/space. However, the reader shouldn't have too many issues following the same steps, using the chain rule, to arrive at:

$$\frac{\partial}{\partial b_{ij}^{(l)}} J\left(w,b,x,y\right) = \delta_i^{(l+1)}$$

So we now know how to perform our original gradient descent problem for neural networks:

$$w_{ij}^{(l)} \leftarrow w_{ij}^{(l)} - \alpha \frac{\partial}{\partial w_{ij}^{(l)}} J\left(w,b,x,y\right)$$

$$b_i^{(l)} \leftarrow b_i^{(l)} - \alpha \frac{\partial}{\partial b_i^{(l)}} J\left(w,b,x,y\right)$$

However, to perform this gradient descent training of the weights, we would have to resort to loops within loops. As previously shown in vectorization in neural networks of this neural network tutorial, performing such calculations in Python using loops is slow for large networks. Therefore, we need to figure out how to vectorize such calculations, which the next section will show.

10.1.7 Vectorization of Backpropagation

To consider how to vectorize the gradient descent calculations in neural networks, let's first look at a naïve vectorized version of the gradient of the cost function (warning: this is not in a correct form yet!):

$$\frac{\partial J}{\partial W^{(l)}} = \delta^{(l+1)} h^{(l)}$$

$$\frac{\partial J}{\partial b^{(l)}} = \delta^{(l+1)}$$

Now, let's look at what element of the aforementioned equations. What does $h^{(l)}$ look like? Pretty simple, just a $s_l \times 1$ vector, where s_l is the number of nodes in layer l. What does the multiplication of $h^{(l)} \delta^{(l+1)}$ look like? Well, because we know that $\alpha \times \dfrac{\partial J}{\partial W^{(l)}}$ must be the same size of the weight matrix $W^{(l)}$, we know that the outcome of $h^{(l)} \delta^{(l+1)}$ must also be the same size as the weight matrix for layer l. In other words, it has to be of size $s_{(l+1)} \times s_l$. We know that $\delta^{(l+1)}$ has the dimension $s_{l+1} \times 1$ and that $h^{(l)}$ has the dimension of $s_{(l)} \times 1$.

The rules of matrix multiplication show that a matrix of dimension ($\mathbf{n} \times \mathbf{m}$) multiplied by a matrix of dimension ($\mathbf{o} \times \mathbf{p}$) will have a product matrix of size ($\mathbf{n} \times \mathbf{p}$). If we perform a straight multiplication between $h^{(l)}$ and $\delta^{(l+1)}$, the number of columns of the first vector (i.e., 1 column) will not equal the number of rows of the second vector (i.e., 3 rows), therefore we can't perform a proper matrix multiplication.

The only way we can get a proper outcome of size $s_{(l+1)} \times s_l$ is by using a matrix transpose. A transpose swaps the dimensions of a matrix around, that is, a ($s_l \times 1$) sized vector becomes a ($1 \times s_l$) sized vector and is denoted by the superscript $\mathbf{T}$. Therefore, we can do the following:

$$\delta^{(l+1)} \times \left(h^{(l)} \right)^T = \left(s_{(l+1)} \times 1 \right) \times (1 \times s_l) = \left(s_{(l+1)} \times s_l \right)$$

As can be observed from the following, by using the transpose operation we can arrive at the desired outcome. A final vectorization that can be performed is during the weighted addition of the errors in the backpropagation step:

$$\delta_j^{(l)} = \left(\sum_{i=1}^{s_{(l+1)}} w_{ij}^{(l)} f'\left(z_j^{(l)}\right) \right) = \left(\left(W^{(l)}\right)^T \delta_i^{(l+1)} \right) \odot \left(f'\left(z_j^{(l)}\right) \right)$$

The $\odot$ symbol in the previous designates an element-by-element multiplication (called the Hadamard product), not a matrix multiplication. Note that the matrix multiplication $\left[\left(W^{(l)}\right)^T \delta^{(l+1)} \right]$ performs the necessary summation of the weights and δ values—the reader can check that this is the case.

10.1.8 Implementing Gradient Descent Step

Now, how do we integrate this new vectorization into the gradient descent steps of our soon-to-be coded algorithm? First, we must look again at the overall cost function we are trying to minimize (not just the sample-by-sample cost function shown in the preceding equation):

$$J\left(w, b\right) = \frac{1}{m} \sum_{z=0}^{m} J\left(W, b, x^z, y^z\right)$$

As we can observe, the total cost function is the mean of all the sample-by-sample cost function calculations. Also remember the gradient descent calculation (showing the element-by-element version along with the vectorized version):

$$w_{ij}^{(l)} \leftarrow \left\{ w_{ij}^{(l)} - \alpha \left[\frac{\partial}{\partial w_{ij}^{(l)}} J(w,b) \right] \right\}$$

$$W^{(l)} \leftarrow \left\{ W^{(l)} - \alpha \left[\frac{\partial}{\partial W^{(l)}} J(w,b) \right] \right\} \leftarrow \left\{ W^{(l)} - \alpha \left[\frac{1}{m} \sum_{z=1}^{m} \frac{\partial}{\partial W^{(l)}} J\left(w,b,x^{(z)},y^{(z)}\right) \right] \right\}$$

So that means as we go along through our training samples or batches, we must have a term that is summing up the partial derivatives of the individual sample cost function calculations. This term will gather up all the values for the mean calculation. Let's call this "summing up" term $\Delta W^{(l)}$. Likewise, the equivalent bias term can be called $\Delta b^{(l)}$. Therefore, at each sample iteration of the final training algorithm, we have to perform the following steps:

$$\Delta W^{(l)} \leftarrow \Delta W^{(l)} + \frac{\partial}{\partial W^{(l)}} J\left(w,b,x^{(z)},y^{(z)}\right) \leftarrow \Delta W^{(l)} + \delta^{(l+1)} \left(h^{(l)}\right)^{T}$$

$$\Delta b^{(l)} \leftarrow \Delta b^{(l)} + \delta^{(l+1)}$$

By performing these operations at each iteration, we slowly build up the previously mentioned sum:

$$\sum_{z=1}^{m} \frac{\partial}{\partial W^{(l)}} J\left(w,b,x^{(z)},y^{(z)}\right)$$

The same is true for b. Once all the samples have been iterated through, and the Δ values have been summed up, we update the weight parameters:

$$W^{(l)} \leftarrow W^{(l)} - \alpha \left[\frac{1}{m} \Delta W^{(l)} \right]$$

$$b^{(l)} \leftarrow b^{(l)} - \alpha \left[\frac{1}{m} \Delta b^{(l)} \right]$$

10.1.9 FINAL GRADIENT DESCENT ALGORITHM

So now, we've finally made it to the point where we can specify the entire backpropagation-based gradient descent training of our neural networks. It has taken quite a few steps to show, but hopefully it has been instructive. The final backpropagation algorithm is as follows:

- Randomly initialize the weights for each layer $W^{(l)}$
- While iterations < iteration limit:
- Set ΔW and Δb to zero.
- For samples 1 to m:
- Perform a feed forward pass through all the n_l layers. Store the activation function outputs $h^{(l)}$.

- Calculate the δ^{nl} value for the output layer.
- Use backpropagation to calculate the $\delta^{(l)}$ values for layers 2 to $(n_l - 1)$
- Update the $\Delta W^{(l)}$ and $\Delta b^{(l)}$ each layer.
- Perform a gradient descent step using:

$$W^{(l)} \leftarrow W^{(l)} - \alpha \left[\frac{1}{m} \Delta W^{(l)} \right]$$

$$b^{(l)} \leftarrow b^{(l)} - \alpha \left[\frac{1}{m} \Delta b^{(l)} \right]$$

As specified in the previous algorithm, we would repeat the gradient descent routine until we are happy that the average cost function has reached a minimum. At this point, our network is trained and (ideally) ready for use. Many neural networks have emerged over the years such as follows:

- Deep Neural Networks
- Modular Neural Networks
- Feedforward Artificial Neural Networks
- Perceptron and Multilayer Perceptron Neural Networks
- Radial-Basis Functions Artificial Neural Networks
- Recurrent Neural Network (RNN)
  - Long Short-Term Memory (LSTM)
  - Gated Recurrent Unit (GRU)
- Conventional Neural Networks (CNN)
- Deep Belief Network (DBN)
- Hierarchical Attention Networks (HAN)
- Combination Techniques
  - Random Multi-Model Deep Learning (RMDL)
  - Stochastic Gradient Descent (SGD) Optimizer
  - Root Square Mean Proportionate (RMSProp)** Step Size Optimizer
  - Adam Optimizer
  - Adaptive Gradient Algorithm (*Adagrad*) Optimizer
  - Adaptive Learning Rate per Dimension (Adadelta) Optimizer
  - Hierarchical Deep Learning for Text (HDLTex)
  - Other Techniques
    - Recurrent Convolutional Neural networks (RCNN)
    - CNN with LSTM (C-LSTM)

10.2 SUMMARY

In this section, we have explained deep learning with basic mathematical formulation that deals with neural networks. Cost and error function, gradient decent, and backpropagation related to the neural network are articulated. The role of hidden layers is described. Finally, the gradient descent algorithm is described. We have listed many kinds of neural networks. Many books that explain all these neural networks in detail are available online. However, the basic treatments that are provided here are sufficient to use those books.

REFERENCE

[1] Thomas, A., "An Introduction to Neural Networks for Beginners," https://www.studocu.com/en-gb/document/edinburgh-napier-university/computational-intelligence/an-introduction-to-neural-networks-for-beginners/3196355.

11 Overfitting and Underfitting

11.1 OVERFITTING

Overfitting occurs when prediction is made based on regularities that appear in the test sample, but which were not present in the training samples. That is, overfitting occurs when the model cannot generalize and fits too closely to the training dataset because the training data size is too small and does not contain enough data samples to accurately represent all possible input data values. Detecting overfitting is a crucial step in the machine learning process. Here's how we can spot it:

- **Validation Set:** Split the data into training and validation sets. If the model performs well on the training set but poorly on the validation set, it's likely overfitting.
- **Learning Curves:** Plot the model's performance on both the training and validation sets over time. If the two curves start to diverge, it's a sign of overfitting.
- **Cross-Validation:** Use cross-validation, where the training data is split multiple times, and the model is evaluated on each split.

11.2 RESOLVING OVERFITTING

11.2.1 LINEAR REGRESSION

The overfitting problem is very complicated to deal with. In short, we can hypothesize that a simpler linear model will be preferred to complex model. To do so, we can impose penalty for complexity, which is termed as regularizer. A typical form for regularizer is as follows: to find a hypothesis h is to minimize with a variable parameter λ: [1]

$$\left(\sum_e error(e,h) \right) + \lambda * regularize(h)$$

Let us consider the simple linear regression equation:

$$\sum_{e \in E_s} \left[\left(Y(e) - \check{Y}^{\bar{w}}(e) \right) \right]^2 = \sum_{e \in E_s} \left[\left(Y(e) - \sum_{i=0}^{n} w_i * X_i(e) \right) \right]^2$$

Now, we will add a loss function and optimize parameter to make the model that can predict the accurate value of Y. The loss function for the linear regression is called as **Residual sum of squares (RSS)**. While using a linear model helps us avoid overfitting, many real-world problems are nonlinear ones. In addition to understanding how to detect overfitting, it is important to understand how to avoid overfitting altogether. The following are several techniques that you can use to prevent overfitting: Regularization, Early Stopping, Cross-validation, Train with more Data, Data Augmentation, and Feature Selection. First, we describe regularization. There are two types of regularization: Ridge Regression and Lasso Regression.

DOI: 10.1201/9781003499466-11

11.2.2 RIDGE REGRESSION

Ridge regression is one of the types of linear regression in which a small amount of bias is introduced so that we can get better long-term predictions. It reduces the complexity of the model and is called L2 Regularizer. In this technique, the cost function is altered by adding the penalty term to it. The amount of bias added to the model is called Ridge Regression penalty. We can calculate it by multiplying with the lambda to the squared weight of each individual feature. The equation for the cost function in ridge regression will be: [1]

$$\sum_{e \in E_s} \left[\left(Y(e) - \check{Y}^{\bar{w}}(e) \right) \right]^2 = \sum_{e \in E_s} \left[\left(Y(e) - \sum_{i=0}^{n} w_i * X_i(e) \right) \right]^2 + \lambda \sum_{i=0}^{n} (w_i)^2$$

In this equation, the penalty term regularizes the coefficients of the model, and hence ridge regression reduces the amplitudes of the coefficients that decreases the complexity of the model. If the values of λ tend to 0, the equation becomes the cost function of the linear regression model. Hence, for the minimum value of λ, the model will resemble the linear regression model. It reduces the complexity of the model by shrinking the coefficients as well as all the features present in the model. A general linear or polynomial regression will fail if there is high collinearity between the independent variables, so to solve such problems, ridge regression can be used. It helps to solve the problems if we have more parameters than samples.

11.2.3 LASSO REGRESSION

Lasso regression is another regularization technique to reduce the complexity of the model. It stands for Least Absolute and Selection Operator. It is like the ridge regression except that the penalty term contains only the absolute weights instead of a square of weights. The equation for the cost function of Lasso regression will be: [1]

$$\sum_{e \in E_s} \left[\left(Y(e) - \check{Y}^{\bar{w}}(e) \right) \right]^2 = \sum_{e \in E_s} \left[\left(Y(e) - \sum_{i=0}^{n} w_i * X_i(e) \right) \right]^2 + \lambda \sum_{i=0}^{n} (|w_i|)^2$$

Since it takes absolute values, it can shrink the slope to 0, whereas ridge regression can only shrink it near to 0. It is also called as L1 regularization. Some of the features in this technique are completely neglected for model evaluation. Hence, the Lasso regression can help us to reduce the overfitting in the model as well as the feature selection.

11.2.4 EARLY STOPPING

In machine learning, early stopping is one of the most widely used regularization techniques to combat the overfitting issue. Early stopping monitors the performance of the model for every epoch on a held-out validation set during the training and terminate the training conditional on the validation performance. This method seeks to pause training before the model starts learning the noise within the model. This approach risks halting the training process too soon, leading to the opposite problem of underfitting. Finding the "sweet spot" between underfitting and overfitting is the goal here. Early stopping is usually noted in deep learning models where the epochs (iterations of model) are halted when the performance against the testing/validation data starts to degrade.

11.2.5 CROSS-VALIDATION

It is a resampling process where the dataset is split into k number of groups, where certain subsets of each group are used for **training** the model and others are used for validating or **testing** the model. The model is then evaluated on how well it did on the **test** data, also called the "unknown data" as this data wasn't used in fitting/training the model. There are numerous cross-validation methods, such as **K-fold** and **Leave-One-Out**. However, the simplest and most well-known method is **hold-out,** where a portion of the data is set aside for later testing.

11.2.6 TRAIN WITH MORE DATA

Expanding the training set to include more data can increase the accuracy of the model by providing more opportunities to parse out the dominant relationship among the input and output variables. That said, this is a more effective method when clean, relevant data is injected into the model. Otherwise, we could just continue to add more complexity to the model, causing it to overfitting.

11.2.7 DATA AUGMENTATION

While it is better to inject clean, relevant data into your training data, sometimes noisy data is added to make a model more stable. However, this method should be done sparingly.

11.2.8 FEATURE SELECTION

When we build a model, we will have several parameters or features that are used to predict a given outcome, but many times, these features can be redundant to others. Feature selection is the process of identifying the most important ones within the training data and then eliminating the irrelevant or redundant ones. This is commonly mistaken for dimensionality reduction, but it is different. However, both methods help to simplify your model to establish the dominant trend in the data.

11.3 UNDERFITTING

When a model has not learned the patterns in the training data well and is unable to generalize well on the new data, it is known as underfitting. Underfitting is the inverse of overfitting, meaning that the statistical model or machine learning algorithm is too simplistic to accurately capture the patterns in the data. An underfit model has poor performance on the training data and will result in unreliable predictions. Underfitting occurs due to high bias and low variance.

11.3.1 RESOLVING UNDERFITTING

There are multiple ways to deal with underfitting:

- **Use Ensemble Methods**: Ensemble methods combine multiple models to create a more accurate prediction. This can help to reduce underfitting by allowing multiple models to work together to capture the underlying patterns in the data.
- **Use Feature Engineering**: Feature engineering involves creating new model features from the existing ones that may be more relevant to the problem at hand. This can help to improve the accuracy of the model and prevent underfitting.
- **Increase the Model Complexity**: If the model is too simple, it may be necessary to increase its complexity by adding more features, increasing the number of parameters, or using a more flexible model. However, this should be done carefully to avoid overfitting.

- **Use a Different Algorithm**: If the current algorithm is not able to capture the patterns in the data, it may be necessary to try a different one. For example, a neural network may be more effective than a linear regression model for some types of data.
- **Increase the Amount of Training Data:** If the model is underfitting due to lack of data, increasing the amount of training data may help. This will allow the model to better capture the underlying patterns in the data.
- **Use Regularization**: Regularization is a technique as described earlier used to prevent overfitting by adding a penalty term to the loss function that discourages large parameter values. It can also be used to prevent underfitting by controlling the complexity of the model.

11.4   SUMMARY

We have discussed both overfitting and underfitting that are the serious problems in ML/DL algorithms computation. First, we have defined overfitting and underfitting and how to detect them. Second, we described how to resolve overfitting and underfitting.

REFERENCE

[1] Poole, D. L. and Mackworth, A. K., "Artificial Intelligence: Foundation of Computational Agents," Second Edition, 2017, Cambridge University Press.

12 Hybrid Learning

In hybrid learning, multiple simple algorithms work together to complement and augment each other. Together they can solve problems that they were not designed to solve alone through extrapolation. Within hybrid learning there are various types of techniques that interact with the data in different ways [1]. The choice of algorithms depends on the problems at hand, and technical expertise is needed to use those algorithms. Semi-supervised, self-supervised, and multi-instance learning are some examples of hybrid learning. Semi-supervised learning work with both labeled and unlabeled data for training and testing, respectively. It is helpful because data changes over time, for example, forecasting is made on the dataset that is not fully labeled. Semi-supervised learning could be used with supervised and unsupervised learning in tandem. The self-supervised learning model combines both unsupervised and supervised learning algorithms and then applies a supervised learning model. It is helpful, for example, in the context of image processing although it might not reveal what the picture itself is about. Multi-instance learning model uses supervised learning models to identify labels for groups of data. The model is trained to recognize attributes of a few pieces of data within a group, and then it predicts labels for future groups based on attributes of some of the data within the new groups. That is, this model is useful where labeling is done for groups or collection of data, rather than the individual members of the group. Multi-instance learning is usually used for large sets of similar data and have a lot of duplicates. Next, we describe some hybrid machine learning algorithms.

12.1 SEMI-SUPERVISED LEARNING

Semi-supervised learning describes a class of algorithms that seek to learn from both unlabeled and labeled samples, typically assumed to be sampled from the same or similar distributions [2]. Approaches differ on what information to gain from the structure of the unlabeled data. There are a wide variety of semi-supervised learning techniques proposed in the literature. For more context, we focus on recent developments based on deep neural networks. The standard protocol for evaluating semi-supervised learning algorithms works as follows:

- Start with a standard labeled dataset.
- Keep only a portion of the labels (say, 10%) on that dataset.
- Treat the rest as unlabeled data.

While this approach may not reflect realistic settings for semi-supervised learning, it remains the standard evaluation protocol, which one follows it for this work. Many of the initial results on semi-supervised learning with deep neural networks were based on generative models, such as denoising autoencoders, variational autoencoders, and generative adversarial networks (GAN). More recently, a line of research showed improved results on standard baselines by adding consistency regularization losses computed on unlabeled data. These consistency regularization losses measure discrepancy between predictions made on perturbed unlabeled data points. Additional improvements have been shown by smoothing predictions before measuring these perturbations. Approaches of these kind include ∏-Model (small machine learning models that don't require a lot of computational power), Temporal Ensembling (combines several models together), Mean Teacher (uses a simple and intuitive model to get better predictions and has the option of utilizing unlabeled data), and Virtual Adversarial Training (an effective regularization technique which has given good results in supervised learning, semi-supervised learning, and unsupervised clustering).

DOI: 10.1201/9781003499466-12

Recently, fast-Stochastic Weight Averaging (fast-SWA) has shown improved results by training with cyclic learning rates and measuring discrepancy with an ensemble of predictions from multiple checkpoints. By minimizing consistency losses, these models implicitly push the decision boundary away from high-density parts of the unlabeled data. This may explain their success on typical image classification datasets, where points in each cluster typically share the same class.

Two additional important approaches for semi-supervised learning which have shown success both in the context of deep neural networks and other types of models are Pseudo-Labeling, where one imputes approximate classes on unlabeled data by making predictions from a model trained only on labeled data, and conditional entropy minimization, where all unlabeled examples are encouraged to make confident predictions on some class. Semi-supervised learning algorithms are typically evaluated on small-scale datasets. We are aware of very few examples in the literature where semi-supervised learning algorithms are evaluated on larger, more challenging datasets.

12.2   SELF-SUPERVISED LEARNING (SSL)

Self-supervised learning (SSL) is a general learning framework that relies on surrogate (pretext) tasks that can be formulated using only unsupervised data. A pretext task is designed in a way that solving it requires learning of a useful image representation. Self-supervised techniques have a variety of applications in a broad range of computer vision topics [2]. Here we employ self-supervised learning techniques that are designed to learn useful visual representations from image databases. These techniques achieve state-of-the-art performance among approaches that learn visual representations from unsupervised images only. In the following, we have provided a non-comprehensive summary of the most important developments in this direction. It is seen that a trained CNN model predicts relative location of two randomly sampled nonoverlapping image patches.

In another case, this idea has been generalized for predicting a permutation of multiple randomly sampled and permuted patches. Beside the patch-based methods, there are self-supervised techniques that employ image-level losses. Among those, it is proposed to use grayscale image colorization as a pretext task. Another example is a pretext task that predicts an angle of the rotation transformation that was applied to an input image. Some techniques go beyond solving surrogate classification tasks and enforce constraints on the representation space. A prominent example is the exemplar loss that encourages the model to learn a representation that is invariant to heavy image augmentations. Another example, the model enforces additivity constraint on visual representation: the sum of representations of all image patches should be close to representation of the whole image. Finally, a learning procedure has been proposed that alternates between clustering images in the representation space and learning a model that assigns images to their clusters.

12.3   MULTI-INSTANCE LEARNING

Multi-Instance Learning (MIL) is a form of weakly supervised learning where training instances are arranged in sets, called bags, and a label is provided for the entire bag, opposed to the instances themselves [3]. This allows to leverage weakly labeled data which is present in many business problems, as labeling data is often costly for applications such as medical imaging, audio/video, text, and time series data. In the standard MIL representation, negative bags are said to contain only negative instances, while positive bags contain at least one positive instance. Positive instances are labeled in the literature as witnesses.

Instant Level vs. Bag Level: In some applications, like object localization in images (in content retrieval, for instance), the objective is not to classify bags but to classify individual instances. The bag label is the presence of that entity in the image. Note that the bag classification performance of a method often is not representative of its instance classification performance. For example, when considering negative bags, a single false positive cause a bag to be misclassified. On the other hand, in positive bags, it does not change the label, which shouldn't affect the loss at bag-level.

Bag Composition: Most existing MIL methods assume that positive and negative instances are sampled independently from a positive and a negative distribution. This is often not the case due to the co-occurrence of several relations:

Intra Bag Similarities: The instances belonging to the same bag share similarities that instances from other bags do not. In computer vision applications, it is likely that all segments share some similarities related to the capture condition (e.g., illumination). Another option is overlapping patches in an extraction process, as represented later.

Instance Co-Occurrence: Instances co-occur in bags when they share a semantic relation. This type of correlation happens when the subject of a picture is more likely to be seen in some environment than in another or when some objects are often found together.

Instance and Bag Structure: In some problems, there is an underlying structure (spatial, temporal, relational, causal) between instances in bags or even between bags. For example, when a bag represents a video sequence—for instance, identifying the frames of a video where a cat appears knowing only there's a cat in that video—all frames or patches are temporally and spatially ordered.

Label Ambiguity: Label noises and different label spaces cause label ambiguities. Some MIL algorithms, especially those working under the standard MIL assumption, rely heavily on the correctness of bag labels. In practice, there are many situations where positive instances may be found in negative bags due to labeling errors or inherent noise. For example, in computer vision applications, it is difficult to guarantee that negative images contain no positive patches: an image showing a house may contain flowers but is unlikely to be annotated as a flower image. Label noise occurs as well when you have different bags with different densities of positive events. For instance, we have an audio recording (R1) of 10 seconds containing only a total of 1 second of the tagged event in it and another audio recording (R2) of the same duration in which the tagged event is present for a total of 5 seconds. R1 is a weaker representation of the event compared to R2. It is possible to extract patches from negative images that fall into this positive region. For example, some patches extracted from the image of a white tiger might fall into another concept region due to being visually like it.

Multiple Instance Learning Models: There are multiple models that can be used for MIL—either at instance or bag-level classification. Bag-level classification can be of two types: Bag of Words approach and Earth Mover Distance Support Vector Machine (EMD-SVM). In Bag of Words approach, a bag can be represented by its instances, using methods such as an image embedding, and determining the frequency of each instance in a bag. A classifier is then trained on this histogram to determine whether a bag is positive or not. The EMD-SVM is a measure of the dissimilarity between two distributions (e.g., via an image embedding as well). Each bag is a distribution of instances, and the EMD is used to create a kernel used in an SVM.

Instant-Space Methods: Alternative applications of SVMs, Multiple Instant SVM (MISVM) were developed for multiple instances learning applications. Classically, SVMs try to determine the maximum margin between instances. For MIL, since the goal is to have at least one instance in a positive bag as positive, the margin is changed so that condition occurs: at least one instance in a positive bag should have a large positive margin. After determining the decision function, the instances' class can be recovered.

Neural Network with Pooling: With a bag-level label, we can have a latent space containing the probability of each segment (using a sequence-based input). By applying a pooling operator (max/average pooling), there's just a single score associated with a bag (see Figure 12.1).

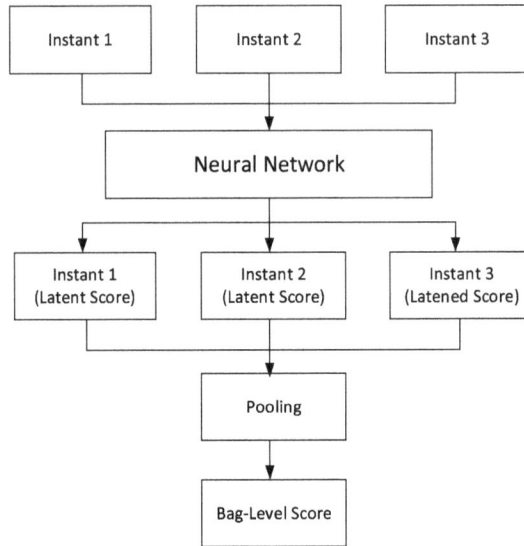

FIGURE 12.1   Multi-Instant Learning.

After training, if you want to do an instance-level prediction, the last pooling layer can be removed. Usually, max pooling is used for classification problems while average pooling is applied to regression problems.

12.4   CONTRASTIVE SELF-SUPERVISED LEARNING

Contrastive self-supervised learning uses both positive and negative examples [4]. Contrastive learning is most notably used for self-supervised learning, a type of unsupervised learning where the label, or supervisory signal, comes from the data itself and classification of data is performed on the basis of similarity and dissimilarity. In the self-supervised setting, contrastive learning allows us to train encoders to learn from massive amounts of unlabeled data. Contrastive learning's loss function minimizes the distance between positive samples while maximizing the distance between negative samples.

12.5   NON-CONTRASTIVE SELF-SUPERVISED LEARNING

The non-contrastive approach relies only on positive sample pairs. For instance, FAIR demonstrated this as the training data containing two versions of a cat picture, the original in color and the other in black and white. There is no inclusion of negative examples, like an unrelated photo of a mountain. The non-contrastive approach uses an extra predictor and a stop-gradient operation that prevents the flow of gradients during forward or reverse-mode automatic differentiation. Two popular non-contrastive methods, Bootstrap Your Own Latent (BYOL) and SimSiam, have proved the need for the predictor and stop-gradient in preventing a representational collapse in the model. Note that BYOL relies on two neural networks, referred to as online and target networks that interact and learn from each other. From an augmented view of an image, the online network is trained to predict the target network representation of the same image under a different augmented view. SimSiam is extraordinarily sensitive to dataset and model size. SimSiam representations undergo partial dimensional collapse if the model is too small relative to the dataset size. Unlike contrastive, the non-contrastive approach is simpler, based on optimizing convolution neural network (CNN) to extract similar feature vectors for similar images. They learn representations by minimizing the distance between two views of the same image. In the cat example, the algorithm would detect characteristics like eyeballs, fur, paws, and whiskers to relate to the cat.

12.6 SELF-SUPERVISED SEMI-SUPERVISED LEARNING (S4L)

Self-Supervised Semi-Supervised Learning (S4L) unifies Self-Supervised and Semi-Supervised Learning approaches yielding a new state-of-the-art result on semi-supervised learning. In this scheme, a semi-supervised training technique that leverages unlabeled data through a combination of novel self-supervised pre-training and established semi-supervised training techniques [5–6]. S4L has been used to construct workflows for data labeling and annotation verification tasks. This unifying approach has demonstrated the effectiveness of workflows over existing methodologies. In addition, it has been experimentally verified that the use of regularizations in S4L improves results further.

12.7 SUMMARY

The scope of hybrid learning is huge. The detailing of this topic needs to be dealt with in a separate book. We have briefly discussed different kinds of hybrid learning: Semi-Supervised Learning, Self-Supervised Learning, Multi-Instant Learning, Contrastive Self-Supervised Learning, Non-Contrastive Self-Supervised Learning, and Self-Supervised Semi-Supervised Learning. However, each of these hybrid learning schemes has to be explained providing full treatment. The key is non-hybrid learning that uses a specific learning algorithm often cannot solve the business problems, for example, mix of label and unlabeled data, mix of balanced and unbalanced dataset, and others. The applications such as integration of all data sources, real-time analysis, and automatic decisions need to be addressed by hybrid learning.

REFERENCES

[1] https://safe.menlosecurity.com/doc/docview/viewer/docNE8E00C6191738517058b75e63a5c2df5fe020d8 e2982225e5dc66fea67fc1020ffab5a7d4eea.
[2] Van-Engelen, J. E. et al., "A Survey on Semi-Supervised Learning," Machine Learning. 109: 373–440, 2020, https://doi.org/10.1007/s10994-019-05855-6.
[3] https://nilg.ai/202105/an-introduction-to-multiple-instance-learning/.
[4] https://ankeshanand.com/blog/2020/01/26/contrative-self-supervised-learning.html.
[5] Bai, H. et al., "Self-Supervised Learning for Data Labeling and Quality Evaluation," November 2022, 2111.10932v1 [cs.CV] 22 Nov 2021.
[6] Zhai, X. et al., "S4L: Self-Supervised Semi-Supervised Learning," IEEE Xplore.

13 Reinforcement Learning

Reinforcement learning (RL) is little different from unsupervised and supervised learning. Reinforcement learning is a machine learning training method based on rewarding desired behaviors and/or punishing undesired ones to its maximum extent. In general, a reinforcement learning agent is able to perceive and interpret its environment, take actions and learn through trial and error. Reinforcement learning differs from supervised learning in that it does not need labeled input/output pairs to be presented and its suboptimal operation to be explicitly corrected. It finds a balance between exploration of unchartered territory and exploitation of the current knowledge that it has already learnt [1].

- Model-Free and Model-Based RL

RL can be divided into two categories: Model-Based and Model-Free. Figure 13.1 shows the schematic view of these two models.

- Model-Free RL
  - Value-Based RL/Q-Learning (e.g., DQN, QR-DQN)
  - Policy-Based RL/Policy Optimization RL (e.g., A2C/A3C)
- Model-Based RL
  - Learn the Model (e.g., World Models, I2A, MBMF, MBVE)
  - Given the Model (e.g., AlphaZero)

Model-Free RL can be based on policy or value (i.e., Q-value). **Value-Based** RL learns the state or state-action value and acts by choosing the best action in the state. Exploration is necessary. **Policy-Based** RL learns directly the stochastic policy function that maps state to action and acts by sampling policy. It is also known as the Policy Optimization RL. **Model-Based RL** learns the model of the world, then plans using the model. Update and replan the model often. This RL can learn the model from the world/environment or a given model can be defined for it. We are providing some examples of Model-Free and Model-Based RL although we describe most of these examples later.

Deep Q-Network (DQN) approximates a state-value function in a Q-Learning framework with a neural network. The deep Q-network (DQN) algorithm is a Model-Free, online, off-policy reinforcement learning method. A DQN agent is a Value-Based reinforcement learning agent that trains

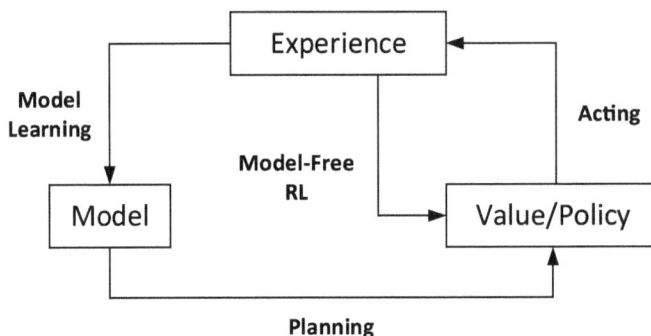

FIGURE 13.1 Schematic View of Model-Free and Model-Based RL.

          DOI: 10.1201/9781003499466-13

a critic to estimate the return or future rewards. DQN is a variant of Q-learning and is also known as Value-Based RL.

Quantile Regression Deep Q-learning network (QR-DQN) RL is a distributional approach to reinforcement learning in which the distribution over returns is modeled explicitly instead of only estimating the mean. This method of learning deals with the value distribution instead of the value function.

The term "actor-critic" is best thought of as a framework or a class of algorithms satisfying the criteria that there exists parameterized *actors* and *critics*. The actor is the policy with parameters which conducts actions in an environment. The critic computes value functions to help assist the actor in learning. These are usually the state value, state-action value, or advantage value. The Advantage Actor Critic (A2C) algorithm combines two types of Reinforcement Learning algorithms (Policy-Based and Value-Based) together. Policy-Based agents directly learn a policy (a probability distribution of actions) mapping input states to output actions. Asynchronous Advantage Actor Critic (A3C) is a policy gradient algorithm in reinforcement learning that maintains a policy and an estimate of the value function. It operates in the forward view and uses a mix of step returns to update both the policy and the value function. The combination of both is termed as the A2C-A3C model.

In Model-Free World Models, a world model is an abstract representation of the spatial or temporal dimensions of our world. A world model can be useful in several ways. One use of a world model is to use their low-dimensional, internal representations for control. The world model with actor-critic behaviors for example automatically learns to compute compact representations of its images that discover useful concepts, such as object positions and learns how these concepts change in response to different actions.

Imagination-Augmented Agents (I2As) is a novel architecture for deep reinforcement learning combining Model-Free and Model-Based aspects. In contrast to most existing Model-Based reinforcement learning and planning methods, which prescribe how a model should be used to arrive at a policy, I2As learn to interpret predictions from a learned environment model to construct implicit plans in arbitrary ways by using the predictions as additional context in deep policy networks.

Model-Based Model-Free (MBMF) Reinforcement Learning aims at bridging the gap between these two Model-Free and Model-Based paradigms that are at the same time data-efficient and cost-savvy. It is done so by learning a probabilistic dynamics model and leveraging it as a prior for the intertwined Model-Free optimization. As a result, this approach can exploit the generality and structure of the dynamics model, but is also capable of ignoring its inevitable inaccuracies, by directly incorporating the evidence provided by the direct observation of the cost. This approach outperforms purely Model-Based and Model-Free approaches, as well as the approach of simply switching from a Model-Based to a Model-Free setting.

The Model-Based value expansion (MBVE), which controls for uncertainty in the model by only allowing imagination to fixed depth. By enabling wider use of learned dynamics models within a Model-Free reinforcement learning algorithm, this model improves value estimation that does not rely on heuristics, which, in turn, reduces the sample complexity of learning.

AlphaZero RL model trains an agent for the game of Go through pure self-play without any human knowledge except the rules of the game. AlphaGo Zero is trained by self-play reinforcement learning. It combines a neural network and Monte Carlo Tree Search in an elegant policy iteration framework to achieve stable learning.

The key is that although reinforcement learning uses Markov decision process (MDP), unlike dynamic programming, it does not assume exact mathematical model of the MDP because for large MDP exact mathematical model becomes infeasible. In general, an RL Agent makes the decisions to optimize a given notion of cumulative rewards. The RL agent modifies or acquires new behaviors and skills incrementally using trial and error experience. RL can be in both offline and inline setting. In an offline setting, the experience is acquired a priori, then it is used as a batch for learning. This is in contrast to the online setting, where data becomes available in a sequential order and is used to progressively update the behavior of the agent.

In both cases, the core learning algorithms are essentially the same, but the main difference is that in an online setting, the agent can influence how it gathers experience so that it is most useful for learning. This is an additional challenge mainly because the agent must deal with the exploration/exploitation dilemma while learning. But learning in the online setting can also be an advantage since the agent is able to gather information specifically on the most interesting part of the environment. For that reason, even when the environment is fully known, RL approaches may provide the most computationally efficient approach in practice as compared to some dynamic programming methods that would be inefficient due to this lack of specificity.

13.1 RL AGENT AND ENVIRONMENT INTERACTION

In fact, RL is feedback-based machine learning that takes suitable actions to maximize reward in a given situation. Figure 13.2 provides RL agent-environment interaction.

Let us consider the Discrete Time Markovian Decision Process (MDP) to define the RL agent-environment interaction. An MDP is a 5-tuple (S,A,T,R,γ), where:

- S is the state space,
- A is the action space,
- T: $S \times A \times S \rightarrow [0,1]$ is the transition function (set of conditional transition probabilities between states),
- P: $S \times S \times A \rightarrow [0,1]$ is the probability that sets the dynamics; for example, $P(s' \mid s,a)$ is the probability of the agent transitioning into state s' given that the agent in state s and does actions a; thus, $\forall s \in S \forall a \in A \sum_{s' \in S} P(s' \mid s,a) = 1$ [1],
- $R : S \times A \times S \rightarrow R$ is the reward function, where $R(s,a,s')$ is a continuous set of possible rewards in a range $R_{max} \in R^+$ (e.g., $[0, R_{max}]$) from ding action a and transitioning to s' from state s; $R(s,a,s') = \sum_{s'} R(s,a,s')*P(s'|s,a)$; note sometimes it is convenient to write $R(s,a,s')$ as $R(s,a)$,
- $\gamma \in [0,1]$ is the discount factor.

The Discrete Time Markovian Decision Process (MDP), $\mathbb{P}$, is defined as follows:

$$\mathbb{P}\left(s_{t+1} \mid s_t, a_t\right) = \mathbb{P}\left(s_{t+1} \mid s_t, a_t, \ldots \ldots, s_0, a_0\right)$$

$$\mathbb{P}\left(r_t \mid s_t, a_t\right) = \mathbb{P}\left(r_t \mid s_t, a_t, \ldots \ldots, s_0, a_0\right)$$

Part of an MDP diagram is shown in Figure 13.3 with state-transition, action, and reward [1].

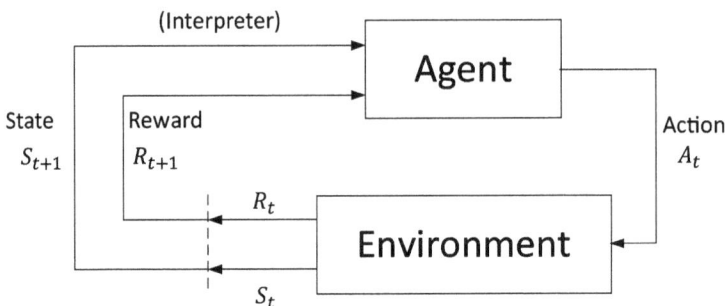

FIGURE 13.2 Interaction between Agent and Environment.

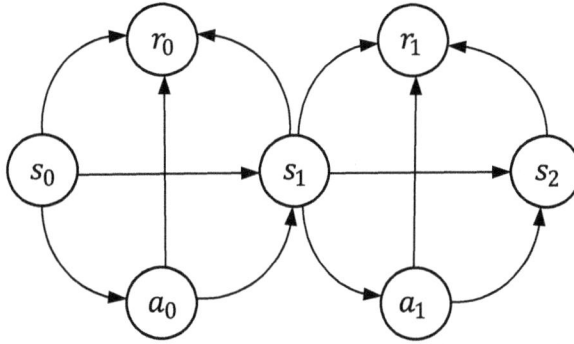

FIGURE 13.3   Partial Markov Decision Process Chain for RL.

An agent compares different sequences of rewards (scalar quantity) for taking actions based on certain policies: $\{r_1, r_2, ..., r_n\}$.

- Total Reward: $V = \sum_{i=1}^{\infty} r_i$
- Average Reward: $V = \lim_{n \to \infty} \{r_1 + r_2 + ... + r_n\} / n$
- Discount Reward/Factor: The discount factor γ essentially determines how much the reinforcement learning agents cares about rewards in the distant future relative to those in the immediate future.

If $\gamma = 1$, the agent will evaluate each of its actions based on the sum of all of its future rewards.

$$V = r_1 + \gamma r_2 + \gamma^2 r_3 + ... + \gamma^n r_n$$

where γ is called the discount factor and its value $0 < \gamma < 10$.

It can be rewritten as follows:

$$V = \sum_{i=1}^{\infty} \gamma^{i-1} r_i = r_1 + \gamma r_2 + \gamma^2 r_3 + ... + \gamma^{i-1} r_i = r_1 + \gamma \left(r_2 + \gamma \left(r_3 + \cdots \right) \right)$$

Suppose V_k is the reward accumulated from time k:

$$V_k = r_k + \gamma \left(r_{k+1} + \gamma \left(r_{k+2} + \cdots \right) \right) = r_k + \gamma V_{k+1}$$

Let us understand the property of V_k:

Suppose $S = 1 + \gamma + \gamma^2 + \gamma^3 + \cdots$, then $S = 1 + \gamma S$; solving for S, we get:

$$S = 1 / (1 - \gamma)$$

With the discounted reward:

- The value of all of the future is at most $1 / (1 - \gamma)$ times as much as the maximum reward.
- The value of all of the future is at least $1 / (1 - \gamma)$ times as much as the minimum reward.

In fact, the discounted reward is referred to as the value.

13.1.1 POLICIES

We are assuming that the Markov decision process is fully observable. A **policy** specifies what the agent should do as a function of the state it is in. Let us consider a stationary policy defined by $\pi : S \rightarrow A$, where π is policy in state $s \in S$ for carrying out an action $a \in A$. However, for a nonstationary policy, the action is a function of the state and time. Given a reward criterion, a policy has an expected value $V^{\pi}(s)$ for every state $s \in S$. Policy π is an optimal policy if there is no policy π' and no state s such that $V^{\pi'}(s) > V^{\pi}(s)$. That is, it is a policy that has a greater or equal expected value at every state than any other policy.

We know that reward γ is a function of an action at a given state at a given instant t: $r(s_t, a_t)$. If we want to maximize our reward for a given time horizon T, we can formulate the expected cumulative reward for parameter θ as follows:

$$\theta^{max} = \max_{\theta} \left[\sum_{t=0}^{T} r(s_t, a_t) \right]$$

We can formulate our problem as follows:

$$R = f(s, a)$$

13.2 OPTIMAL STATIONARY POLICY FOR INFINITE HORIZON PROBLEMS

For infinite horizon problems, a stationary MDP always has an optimal stationary policy. However, for finite-stage problems, a nonstationary policy might be better than all stationary policies. For example, if the agent had to stop at n, for the last decision in some state, the agent would act to get the largest immediate reward without considering future actions. However, for earlier decisions, it may decide to get a lower reward immediately to obtain a larger reward later.

13.3 VALUE OF A POLICY

Consider how to compute the expected value, using the discounted reward of a policy, given a discount factor of γ. The value is defined in terms of two interrelated functions: $V^{\pi}(s) =$ Expected value of following policy π in state s (it is does NOT consider action a into account directly). $Q^{\pi}(s, a) =$ (means **Q-value** of policy π; it does act a directly into account). Expected value, starting in state s of doing action a, then the following policy π that is called **Q-value** of policy π.

Q^{π} and V^{π} are defined recursively in terms of each other. If the agent is in state s, and performs action a, and arrives at state s', it gets the immediate reward of $R(s,a,s')$ plus the discounted future reward $\gamma V^{\pi}(s')$. When the agent is planning, it does not know the actual resulting state, so it uses the expected value, averaged over the possible resulting states as follows: Immediate present reward: $P(s'|s,a) * R(s,a,s') +$ (Discounted future reward: $P(s'|s,a) * \gamma V^{\pi}(s')$):

$$Q^{\pi}(s,a) = \sum_{s'} P(s'|s,a) * \left\{ R(s,a,s') + \gamma V^{\pi}(s') \right\}$$

$$= \sum_{s'} P(s'|s,a) * R(s,a,s') + \sum_{s'} P(s'|s,a) * \gamma V^{\pi}(s')$$

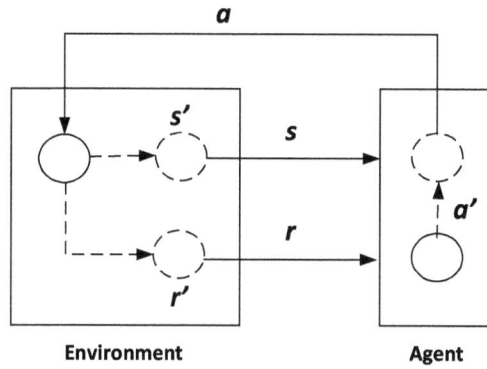

FIGURE 13.4 RL Agent-Environment with State/Action Changes.

Let us define $R(s,a) \triangleq \sum_{s'} P(s'|s,a) * R(s,a,s')$, so we have:

$$Q^{\pi}(s,a) = R(s,a) + \gamma * \sum_{s'} P(s'|s,a) * V^{\pi}(s')$$

$V^{\pi}(s)$ is obtained by doing the action specified by π and then following π:

$$V^{\pi}(s) = Q^{\pi}(s, \pi(s))$$

13.4 Q-TABLE

Q-Table represents the values of an action in (state, action) pair for exploration (Figure 13.5a) while reward table represents the value of the reward for the same (state, action) pair for exploration (Figure 13.5b). Reward matrix contains the positive or negative reward for each state and action

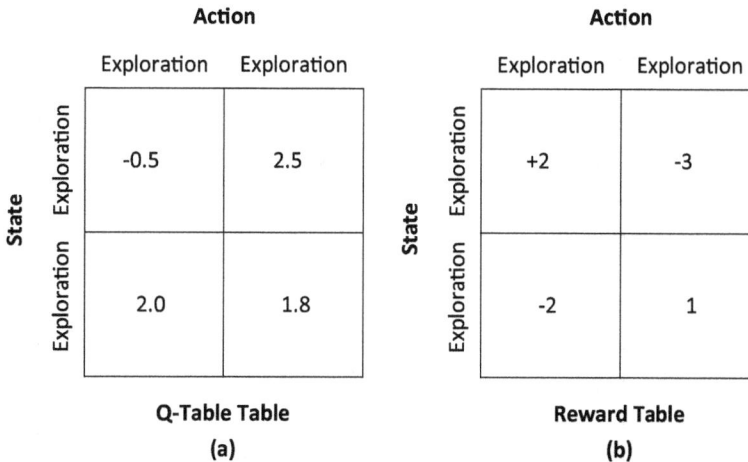

FIGURE 13.5 Q-Table and Reward Table.

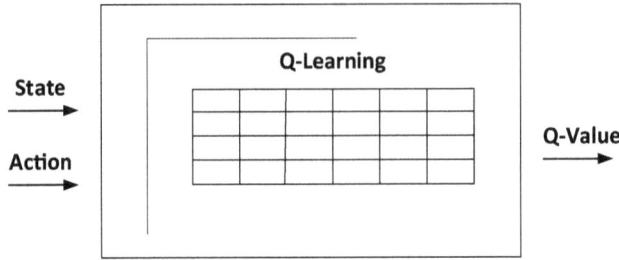

FIGURE 13.6   Q-Learning Table.

couples (Figure 13.5b) while values inside the Q-Table matrix describe the result of choices that are taken by the RL agent in the previous episodes. The relevant reward-Table and Q-Table are shown in Figure 13.5. The proposed mechanism helps in finding the most appropriate decision that can be made during exploring and exploiting phases.

Q Learning (Figure 13.6) is a Model-Free reinforcement learning algorithm and builds a Q-table of State-Action values, with dimension (s,a) where s is the number of states, and a is the number of actions. It can handle problems with stochastic transitions and rewards without requiring adaptations. Fundamentally, a Q-table maps state and action pairs to a Q-value. In practice, the number of states and actions could be huge, making it computationally intractable to build a table.

Q-learning finds an optimal policy in the sense of maximizing the expected value of the total reward over all successive steps, starting from the current state using (finite) Markov decision process (MDP). It can identify an optimal action-selection policy with expected rewards for an action taken in each state given infinite exploration and a partly random policy.

13.5   VALUE OF AN OPTIMAL POLICY

Let $Q^*\left(s,a\right)$ be the expected value of doing a in state s and then following the optimal policy, and $V^*(s)$ be the expected value of following an optimal policy from state s.

Q^* can be defined analogously to Q^π:

$$Q^*\left(s,a\right)=\sum_{s'}P\left(s'\,|\,s,a\right)*\left\{R\left(s,a,s'\right)+\gamma V^*\left(s'\right)\right\}=R\left(s,a\right)+\gamma*\sum_{s'}P\left(s'\,|\,s,a\right)*V^*\left(s'\right)$$

$V^*(s)$ is obtained by performing the action that gives the best value in each state:

$$V^*\left(s\right)=\max_a Q^*\left(s,a\right)$$

An optimal policy π^* is one of the policies that give the best value for each state:

$$\pi^*\left(s\right)=\arg\max_a Q^*\left(s,a\right)$$

where $\arg\max_a Q^*\left(s,a\right)$ is a function of state s, and its value is an action a that results in the maximum value of $Q^*\left(s,a\right)$.

The system is fully observable in an MDP, which means that the observation is the same as the state of the environment: $S_t=s'_t$. At each time-step t, the probability of moving to s_{t+1} is given by the state transition function $T\left(s_t,a_t,s_{t+1}\right)$, and the reward is given by a bounded reward function $R\left(s_t,a_t,s_{t+1}\right)\in R$. This is illustrated in Figure 13.7.

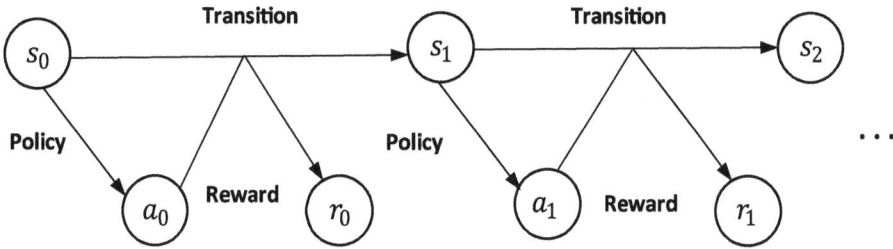

FIGURE 13.7   MDP Changes in State and Reward with Action of the Agent.

13.6   POLICY CATEGORIES

A policy defines how an agent selects actions. Policies can be categorized under the criterion of being either stationary or nonstationary. A nonstationary policy depends on the time-step and is useful for the finite-horizon context, where the cumulative rewards that the agent seeks to optimize are limited to a finite number of future time-steps. We are assuming infinite horizons, and the policies are stationary for analyzing deep RL. Policy criteria can also be categorized either deterministic or stochastic: in the deterministic case, the policy is described by $\pi(s): S \to A$.

In the stochastic case, the policy is described by:

$$\pi(s+a): S \times A \to [0,1]$$

where $\pi(s+a)$ denotes the probability that action a may be chosen in state s. One way to solve RL algorithms [1] is to treat this as an optimization problem with the aim of selecting a policy that maximizes the expected reward collected. The policy can be repetitively evaluated in the environment and iteratively improved. This process is called an Evolutionary Algorithm (EA). This is often combined with Genetic Algorithms (i.e., closed to biological analogy of gene mutation). EA has its own pros and cons (e.g., can be very sensitive to the way policy is represented).

An alternative that we will be pursuing is as follows:

- Learn after every action.
- Learn a component of a policy rather the policy.
- Learning can be linear, polynomial time and space complexity in the number of states, rather than exponential in number of states.

13.7   TEMPORAL DIFFERENCE

Temporal difference (TD) learning is likely the most core concept in reinforcement learning. Temporal difference learning, as the name suggests, focuses on the differences the agent experiences in time. The methods aim to, for some policy π, provide and update some estimate V for the value of the policy $v\pi$ for all states or state-action pairs, updating as the agent experiences them.

The most basic method for TD learning is the $TD(0)$ method. Temporal-Difference $TD(0)$ learning updates the estimated value of a state V for policy based on the reward the agent received and the value of the state it transitioned to. Specifically, if our agent is in a current state s_t, takes the action at and receives the reward r_t, then we update our estimate of V:

$$V(s_t) \leftarrow V(s_t) + \alpha \left[r_{t+1} + \gamma V(s_{t+1}) - V(s_t) \right]$$

where $r_{t+1} + \gamma V\left(s_{t+1}\right)$ is the TD target.

Temporal Difference learning is just trying to estimate the value function $v_\pi\left(s_t\right)$, as an estimate of how much the agent wants to be in certain state, which we repeatedly improve via the reward outcome and the current estimate of $v_\pi\left(s_{t+1}\right)$. This way, the estimate of the current state relies on the estimates of all future states, so information slowly trickles down over many runs through the chain.

13.7.1 TEMPORAL DIFFERENCES (TD)

A simple example of TD will clarify further [1]. To know how RL works, an example in TD error can be examined from a sequence of numerical values: $\left\{v_1, v_2, v_3, \ldots, v_k\right\}$. Let A_k be the running average as follows:

$$A_k = \left\{v_1 + v_2 + v_3, \ldots + v_k\right\} / k$$

Thus,

$$k * A_k = \left\{v_1 * v_2 * v_3, \ldots + v_{k-1} + v_k\right\} = \left(k-1\right) * A_{k-1} + v_k$$

$$A_k = \left(1 - \frac{1}{k}\right) * k * A_{k-1} + \frac{v_k}{k} = \left(1 - \alpha_k\right) * A_{k-1} + \alpha_k * v_k$$

where, $\alpha_k = 1 / k$; then after manipulating:

$$A_k = A_{k-1} + \alpha_k * \left(v_k - A_{k-1}\right)$$

Temporal Difference (TD) Error $\equiv \left(v_k - A_{k-1}\right)$; it specifies how different the new value v_k is from the old prediction, A_{k-1}. It shows that we need to update the old estimate A_{k-1} by TD Error to get the new estimate A_k. This analysis assumes that each number has the equal weight. In practice, the equal weight for each number may have different weights. For example, stock market share values go down in the short term but may tend to increase slowly. So prediction needs to consider different weights of different numbers. In RL, the values are estimates of the effects of actions; more recent values are more accurate than earlier values because the agent is learning, and so they should be weighted more. We can use parameter α_k in such a way it produces more accurate prediction while the solution converges.

$TD(\lambda)$ is a learning algorithm where parameter λ refers to the trace decay parameter with $0 \leq \lambda \leq 1$. Higher settings lead to longer lasting traces with a larger portion of credit from the reward and can be given to more distance states and actions when λ is higher. Note that when $\lambda = 1$, it produces parallel learning to Monte Carlo RL algorithms.

13.8 POLICY GRADIENT

Policy gradient is to search for an optimal policy that maximizes the expected cumulative long-term reward learning a stochastic policy while value function cannot. Policy gradient methods can be used either Model-Free or Model-Based as they are a generic formulation. When RL is applied to real-life examples like robotics and control systems, the uncertain state information causes serious problems and partially observable MDP method needs to be applied for computation which often results in excessive computational demands. These methods provide no guarantee in convergence guarantees, and there exist even divergence examples. Moreover, continuous states and actions in high-dimensional spaces cannot be treated by most off-the-shelf RL algorithms.

Policy gradient methods do not suffer from these problems in the same way. For example, uncertainty in the state might degrade the performance of the policy (if no additional state estimator is being used), but the optimization techniques for the policy do not need to be changed. Continuous states and actions can be dealt with in the same way as discrete ones while, in addition, the learning performance is often increased. Convergence at least to a local optimum is guaranteed.

Let us assume the control system in a discrete-time manner, and we will denote the current time-step by k. In order to take possible stochasticity of the plant into account, we denote it using a probability distribution $x_{k+1} \sim p(x_{k+1} \mid x_k, a_k)$ as model, where $a_k \in \mathbb{R}^M$ denotes the current action, and $x_k, x_{k+1} \in \mathbb{R}^N$ denotes the current and next state, respectively. We furthermore assume that actions are generated by a policy $a_k \sim \pi_\theta(a_k \mid x_k)$ which is modeled as a probability distribution to incorporate exploratory actions; for some special problems, the optimal solution to a control problem is actually a stochastic controller. The policy is assumed to be parameterized by K policy parameters $\theta \in \mathbb{R}^k$. The sequence of states and actions forms a trajectory denoted by $\tau = (x_{0:T-1}, a_{0:T-1})$, where T denotes the horizon which can be infinite. In this chapter, we will use the words trajectory, history, trial, or roll-out interchangeably. At each instant of time, the learning system receives a reward denoted by $r_k = (x_k, a_k) \in \mathbb{R}$.

The general goal of policy optimization in reinforcement learning is to optimize the policy parameters $\theta \in \mathbb{R}^M$ so that the expected return:

$$J(\theta) = E\left\{ \sum_{k=0}^{T-1} a_k r_k \mid \pi_\theta \right\}$$

is optimized where a_k denote time-step dependent weighting factors, often set to $a_k = \gamma^k$ for discounted reinforcement learning, (where γ is in $[0,1]$) or $a_k = 1/T$ for the average reward case. We can rewrite as follows:

$$J(\theta) = \sum_{t=i}^{T-1} P(s_t, a_t \mid \tau) r_{t+1}$$

where i is the arbitrary starting point and $P(s_t, a_t \mid \tau)$ is the probability of occurrence of (s_t, a_t) given the trajectory τ. For real-world applications, we require that any change to the policy parameterization has to be smooth as drastic changes can be hazardous for the actor as well as useful initializations of the policy based on domain knowledge would otherwise vanish after a single update step. For these reasons, policy gradient methods which follow the steepest descent on the expected return are the method of choice. These methods update the policy parameterization according to the gradient update rule:

$$\theta_{t+1} = \theta_t + \eta_t \nabla_\theta J \mid_{\theta=\theta_t}$$

where $\eta_t \in \mathbb{R}^+$ denotes a learning rate and $t \in \{0,1,2,...\}$ the current update number.

The time-step k and update number h are two different variables. In actor-critic-based policy gradient methods, the frequency of updates of t can be nearly as high as of k. However, in most episodic methods, the policy update t will be significantly less frequent. Here, cut-off allows updates before the end of the episode (for $a_k = \gamma^k$, it is obvious that there comes a point where any future reward becomes irrelevant; a generically good cut-off point). If the gradient estimate is unbiased and learning rates fulfill $\sum_{t=0}^{\infty} \eta_t > 0$ and $\sum_{t=0}^{\infty} \eta_t^2 > 0$, the learning process is guaranteed to converge at least to a local minimum.

The main problem in policy gradient methods is to obtain a good estimator of the policy gradient $\nabla_\theta J \mid_{\theta=\theta_t}$. In robotics and control, people have traditionally used deterministic Model-Based methods for obtaining the gradient [2]. However, it is not always possible to be able to solve the model with every detail for autonomous system. Therefore, we need to estimate the policy gradient simply from data generated during the execution of a task, that is, without the need for a model.

Differentiating both sides with respect to policy parameter θ:

$$\nabla_\theta J(\theta) = \sum_{t=i}^{T-1} \nabla_\theta \left[P(s_t, a_t \mid \tau) r_{t+1} \right] = \sum_{t=i}^{T-1} P(s_t, a_t \mid \tau) \frac{\nabla_\theta \left[P(s_t, a_t \mid \tau) \right]}{P(s_t, a_t \mid \tau)} r_{t+1}$$

$$= \sum_{t=i}^{T-1} P(s_t, a_t \mid \tau) \nabla_\theta \log P(s_t, a_t \mid \tau) r_{t+1} == E_\tau \left[\sum_{t=0}^{T-1} \nabla_\theta \log \pi_\theta (a_t \mid s_t) G_t \right] \quad \text{where } G_t \text{ is}$$

the expected cumulative reward.

So, we have:

$$\nabla_\theta J(\theta) = E_\tau \left[\sum_{t=0}^{T-1} \nabla_\theta \log \pi_\theta (a_t \mid s_t) G_t \right] E$$

After analyzing this equation, we can derive the policy gradient with approximation:

$$\nabla_\theta J(\theta) = \left[\sum_{t=0}^{T-1} \nabla_\theta \log \pi_\theta (a_t \mid s_t) G_t \right]$$

13.9 EXPECTED RETURN

In RL, an episode is considered as the agent-environment interaction from initial to final states. Let us define the overall reward be G as the expected reward over an entire episode that an interaction between the agent and the environment. The expected return of a policy over an entire episode is defined, for instance, the state value function $V^\pi (s)$ is provided by $E^\pi \left[G_t \mid s_t = s \right]$ in state s and following policy π. In another example, if the rewards are discounted by a factor $\gamma \in [1,0]$ per time-step, the return at time is $\sum_{t'=t}^{T} \gamma^{t'-t} r_{t'}$, where T is the step when episode terminates, the expected return can be calculated as $\left(\sum_{t'=t}^{T} \gamma^{t'-t} r_{t'} \right) / T$.

13.10 V-VALUE FUNCTION

Throughout this survey, we consider the case of an RL agent whose goal is to find a policy $\pi (s, a) \in \Pi$, so as to optimize an expected return $V^\pi (s): S \rightarrow R$ (also called V-value function) such that:

$$V^\pi (s) = E \left[\sum_{k=0}^{\infty} \gamma^k r_{t+k} \mid s_t = s, \pi \right]$$

where,

- $r_t = E_{a \sim \pi(s_t,.)} R(s_t, a, s_{t+1})$
- $P(s_{t+1} \mid s_{t+1}, a_t) = T(s_t, a_t, s_{t+1})$ with $a \sim \pi(s_t,.)$

Note that *V*-value function is NOT directly related to *A*-action function. From the definition of the expected return, the optimal expected return can be defined as:

$$V^* (s) = \max_{\pi \in \Pi} V^\pi (s)$$

13.11   Q-VALUE FUNCTION

In addition to the V-value function, a few other functions of interest can be introduced. The Q-value function $Q^\pi(s,a): S \times A \to R$ is defined as follows:

$$Q^\pi(s,a) = E\left[\sum_{k=0}^{\infty} \gamma^k r_{t+k} \mid s_t = s, a_t = a, \pi\right]$$

This equation can be rewritten recursively in the case of an MDP using Bellman's equation: Immediate present reward $- P(s'\mid s,a) * R(s,a,s')) +$ Discounted future reward $-P(s'\mid s,a) * \gamma V^\pi(s'))$.
 Note that,

$$P(s'\mid s,a) = T(s,a,s')$$

$$Q^\pi(s,a) = \sum_{s'\in S} T(s,a,s')\left(R(s,a,s') + \gamma Q^\pi(s', a = \pi(s'))\right)$$

The optimal Q-value function $Q^\pi(s,a)$ can also be defined as:

$$Q^*(s,a) = \max_{a\in A} Q^\pi(s,a)$$

Note that Q-value function is directly related to A-action function. The particularity of the Q-value function as compared to the V-value function is that the optimal policy can be obtained directly from $Q^*(s,a)$:

$$\pi^*(s) = \underset{a\in A}{\text{argmax}}\, Q^*(s,a)$$

The optimal V-value function $V^*(s)$ is the expected discounted reward when in each state s while following the policy π^* thereafter. The optimal Q value $Q^*(s,a)$ is the expected discounted return when in each state s and for a given action a while following the policy π^* thereafter. It is also possible to define the Advantage Function:

$$A^\pi(s,a) = Q^\pi(s,a) - V^\pi(s)$$

This quantity describes how good the action a is, as compared to the expected return when directly following policy π. Note that one straightforward way to obtain estimates of either $V^\pi(s)$, $Q^\pi(s)$ or A^π is to use Monte Carlo methods, i.e., defining an estimate by performing several simulations from s while following policy π. In practice, we will see that this may not be possible in the case of limited data. In addition, even when it is possible, we will see that other methods should usually be preferred for computational efficiency.

13.12   FITTED Q-LEARNING

The fitted Q-iteration algorithm is a batch mode reinforcement learning algorithm which yields an approximation of the Q-function corresponding to an infinite horizon optimal control problem with discounted rewards, by iteratively extending the optimization horizon: experiences are gathered in a given dataset D in the form of tuples $\langle s,a,r,s'\rangle$, where the state at the next time-step s' is drawn from $T(s,a,\cdot)$, and the reward r is given by $R(s,a,s')$. In fitted Q-learning (Gordon, 1996), the algorithm starts with some random initialization of the Q-values $Q(s,a;\theta)$, where θ_0 refers to the initial parameters (usually such that the initial Q values should be relatively close to 0 so as to avoid slow learning). Then, an approximation of the Q-values at the *kth* iteration $Q(s, a; \theta_k)$ is updated toward the target value:

$$Y_k^Q = r + \gamma * \max_{a' \in A} Q(s', a'; \theta_k)$$

where θ_k refers to some parameters that define the Q-values at the *kth* iteration.

13.13   DEEP Q-NETWORKS

The memory and computation required for the Q-value algorithm are too high. The deep network Q-Learning function is an approximator for reducing states for algorithms using deep neural network (DNN). This learning algorithm is called Deep Q-Network (DQN). The key idea in this development has thus been to use deep neural networks to represent the Q-network as represented in Figure 13.8 while Figure 13.9 depicts the simplified view of DQN.

Previous attempts at bringing deep neural networks into reinforcement learning were primarily unsuccessful due to instabilities. Deep neural networks are prone to overfitting in reinforcement learning models, which disables them from being generalized. However, DQN algorithms address these instabilities by using two Q-networks: Estimator (Current) Q-Network and Target Q-network. The idea of the Target network is to compute the "target" using the parameters from the Target network that have not been updated for some number of time-steps. In this way, the targets do not "move" during training. Every time-steps, the Target Network is synchronized with the current

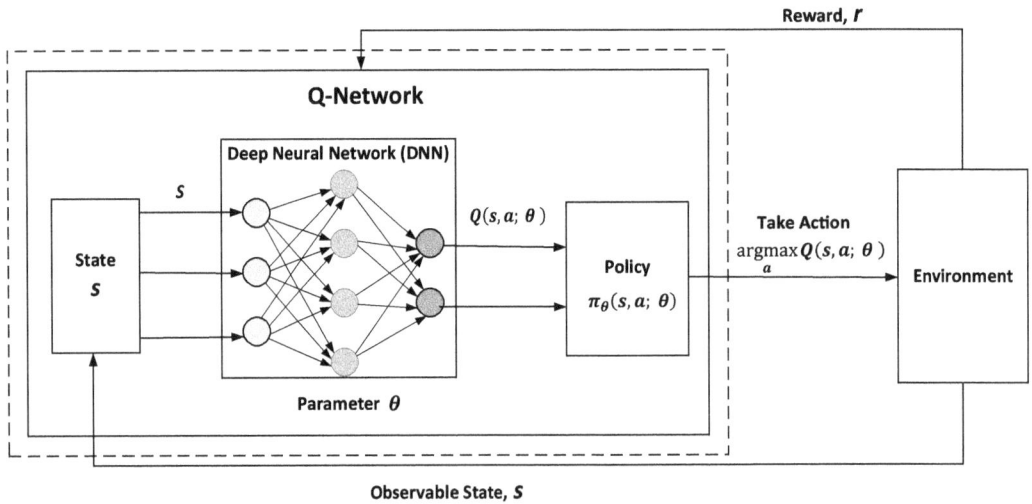

FIGURE 13.8   Schematic View of Deep Q-Network.

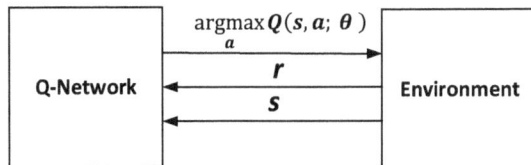

FIGURE 13.9   Simplified Representation of Q-Network using the Deep Neural Network.

network. In addition, DQN uses Replay Memory buffer to have diverse and de-correlated training data by storing all of the agent's experiences and randomly sampling and replaying the experiences. Moreover, a parameter server is needed to store the latest updated parameters, such as gradient descent keeping global states of the parameters synchronized with both target and estimator (current) network.

Note that the experience replay memory of the DQN stores its episode in memory for off-policy learning. In DQN, Experience Replay, as explained earlier, is used to make the network updates more stable. At each time-step of data collection, the transitions are added to a circular buffer called the *replay buffer*. Then during training, instead of using just the latest transition to compute the loss and its gradient, DQN computes them using a mini batch of transitions sampled from the replay buffer. This has two advantages: better data efficiency by reusing each transition in many updates and better stability using uncorrelated transitions in a batch. Note that a Q-network itself can also keep small buffers to store experiences of episodes.

Additionally, the Q-Network is usually optimized toward a frozen target network that is periodically updated with the latest weight (w) every η steps (where η is a hyperparameter termed as learning rate). The latter makes training more stable by preventing short-term oscillations from a moving target. The former tackles autocorrelation that would occur from online learning and having a replay memory makes the problem more like a supervised learning problem. Figure 13.10 shows the high-level complete schematic picture of the overall DQN RL system [3] that does not include the parameter server.

A parameter server (Figure 13.11) is needed for scalable design of the DQN when many learners/workers compute the gradients. The parameter server maintains the globally shared model parameters and aggregate updates from learner/workers. Each learner/worker node pulls the latest model parameters from the server, computes all gradients, and pushes them back for updating. As this approach generally reduces the amount of inter-node communication, it may provide for considerably reduced training time.

A central task is to design a synchronization policy which coordinates the execution progress of all learners (or workers). This synchronization policy determines in each step, that is, whenever a gradient is pushed by some learners/workers, the state ("run" or "wait") of each learner/worker, until

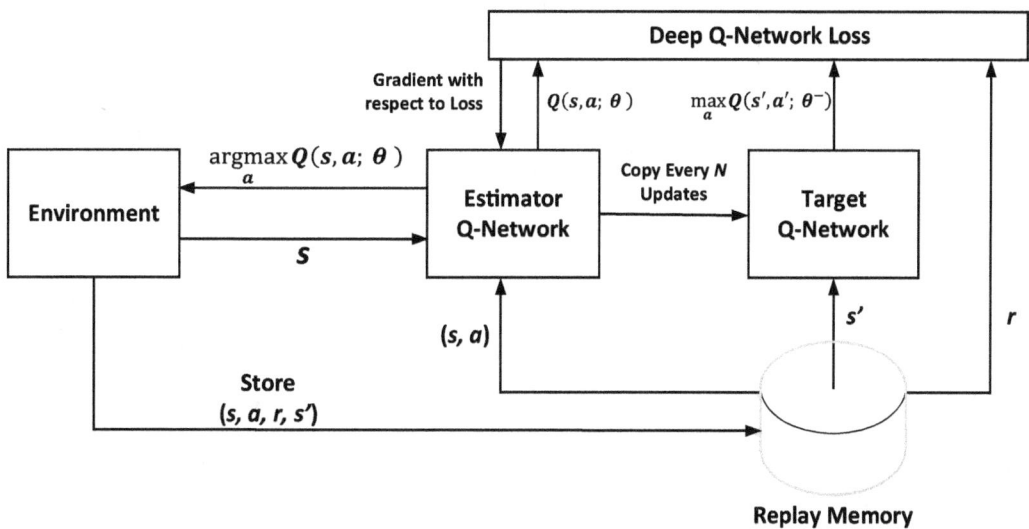

FIGURE 13.10 Deep Q-Network with Estimator Q-Network, Target Q-Network, and Replay Memory.

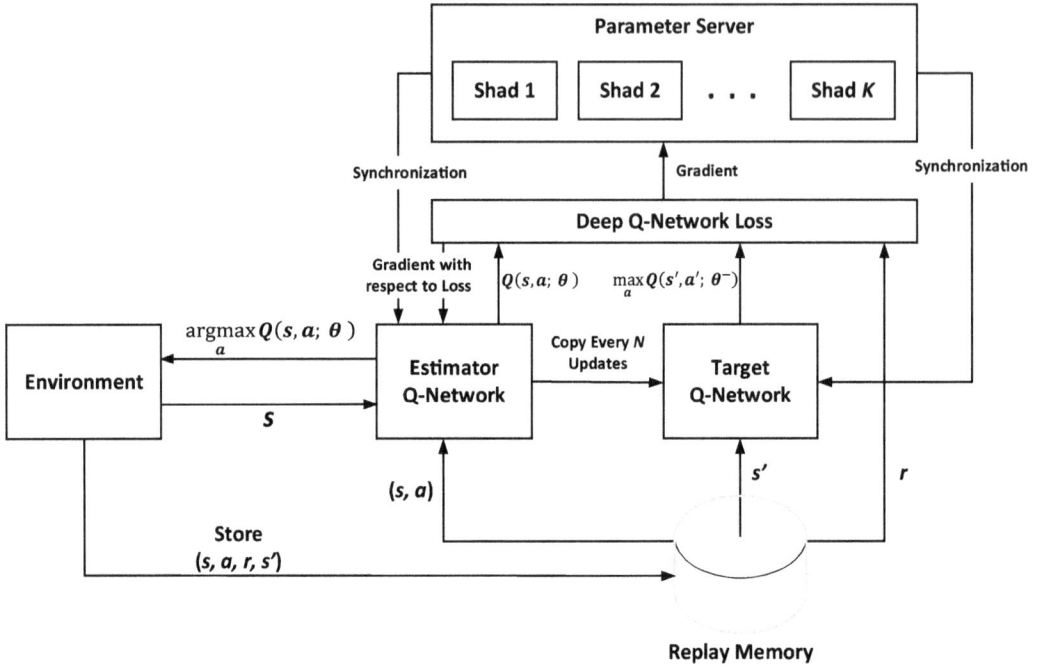

FIGURE 13.11 Deep Q-Network with Estimator Q-Network, Target Q-Network, Replay Memory, and Parameter Server.

the next update arrives at the parameter server. Thus, it directly determines the overall training time. In this context, we have added the parameters server with multiple shads that provide the scalable high-performance algorithms and data structures in the DQN computing architecture.

13.14 MASSIVE PARALLEL DQN ARCHITECTURE

The General Reinforcement Learning Architecture (GORILA) [3], a massively distributed reinforcement learning is depicted in Figure 13.11. The difference between Figures 13.11 and 13.12 is this: the architecture in Figure 13.11 contains only one actor; however, Figure 13.12 has massive numbers of actors.

The learner contains a replica of the Q-network, and its job is to compute desired changes to the parameters of the Q-network. The learner reads samples of the stored experience from the experience replay memory and updates the value function or policy according to a given RL algorithm. For example, a variant of asynchronous stochastic gradient descent (ASGD) algorithms can be used, and these require updates to the parameters of the Q-network.

Specifically, for k-th update, a minibatch of experience tuples $e_t = \left(s_t, a_t, r_t, s_{t+1}\right)$ is sampled from either a local or global experience replay memory $D_t = \{e_1, \ldots, e_t\}$. DQN maintains two separate Q-networks $Q(s, a; \theta)$ and $Q\left(s, a; \theta^-\right)$ with current parameters θ and old parameters θ^-, respectively. The current parameters θ may be updated many times per time-step and are copied into old parameters θ^- after, say, N iterations. At every updated iteration i the current parameters θ are updated so as to minimize the mean-squared Bellman error with respect to old parameters θ^-, by optimizing the following DQN loss function:

$$L_i\left(\theta_i\right) = \mathrm{E}\left[\left(r + \gamma \max_{a'} Q\left(s', a'; \theta^-\right) - Q\left(s, a; \theta_i\right)\right)^2\right]$$

Synchronize with Every Global Steps

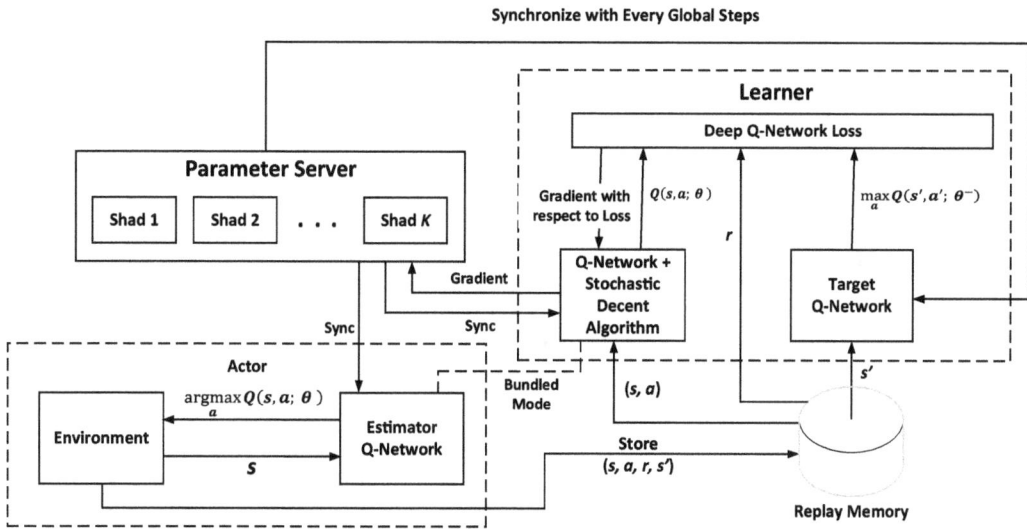

FIGURE 13.12   High-Level Complete Schematic Picture of RL System with Parameter Server.

For each update i, a tuple of experience $(s,a,r,s') \sim U(D)$ or a minibatch, the current parameter θ is updated by a stochastic gradient descent algorithm. Specifically, θ is adjusted in the direction of the sample gradient g_i of the loss with respect to θ:

$$g_i = \left(r + \gamma \max_{a'} Q\left(s',a';\theta^- \right) - Q\left(s,a;\theta_i \right) \right) \nabla_{\theta_i} Q\left(s,a;\theta \right)$$

Finally, actions are selected at each time-step t by an ε-greedy behavior with respect to the current Q-network $Q(s,a;\theta)$.

13.15   DDQN: DOUBLE DEEP Q-NETWORKS

A double deep Q-network (DDQN) uses two Estimator DQN and one Target DQN (Figure 13.13). The Q-value is determined using the expected value of the two Estimator DQN networks. DDQN is an improvement over the single DQN. These overestimations result from a positive bias that is introduced because Q-learning uses the maximum action value as an approximation for the maximum expected action value. DDQN reduces overestimations by decomposing the maximum operation in the target into action selection and action evaluation. Although not fully decoupled, the target network in the DQN architecture provides a natural candidate for the second value function, without having to introduce additional networks. The greedy policy is used according to the online network but using the target network to estimate its value.

13.15.1   BACKGROUND

To solve sequential decision problems, we can learn estimates for the optimal value of each action, defined as the expected sum of future rewards when taking that action and following the optimal policy thereafter. Under a given policy π, the true value of an action in a state s is:

$$Q_\pi \left(s,a \right) \equiv E\left[R_1 + \gamma R_2 + \ldots \mid S_0 = s, A_0 = a, \pi \right]$$

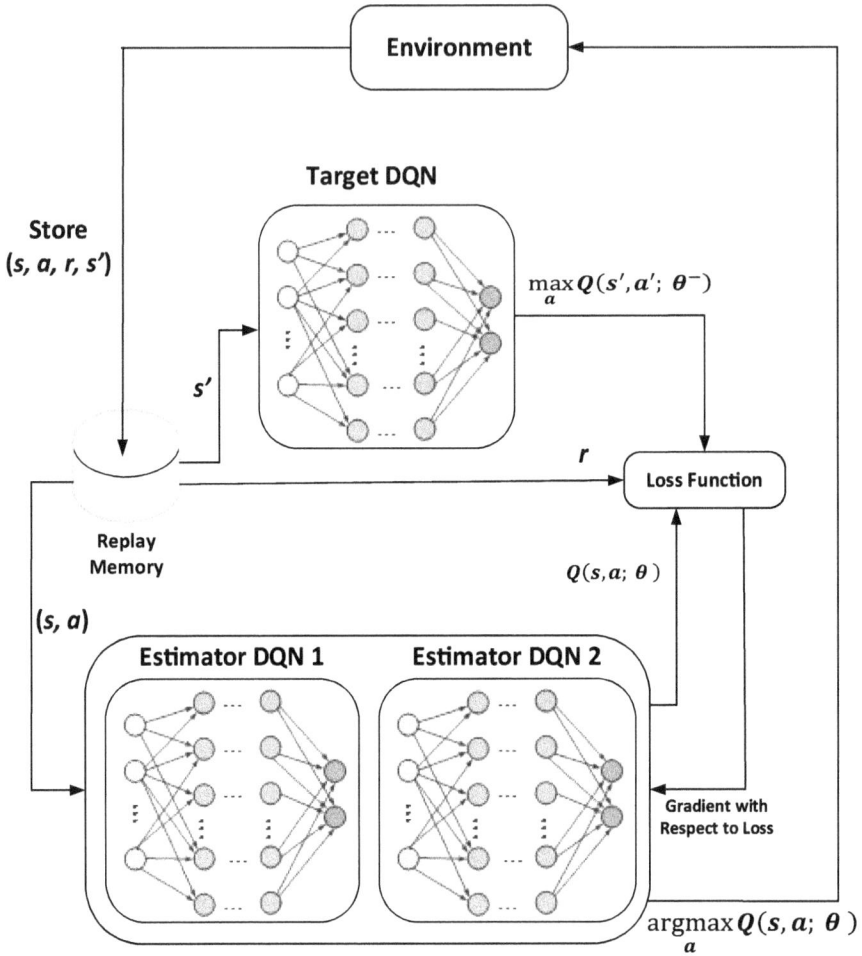

FIGURE 13.13 Double DQN.

where $\gamma \in [0,1]$ is a discount factor that trades off the importance of immediate and later rewards. The optimal value is then $Q_*(s,a) = \max_\pi Q_\pi(s,a)$. An optimal policy is easily derived from the optimal values by selecting the highest-valued action in each state.

Estimates for the optimal action values can be learned using Q-learning, a form of temporal difference learning [4]. Most interesting problems are too large to learn all action values in all states separately. Instead, we can learn a parameterized value function $Q(s,a;\theta_t)$. The standard Q-learning update for the parameters after acting A_t in state S_t and observing the immediate reward R_{t+1} and resulting state S_{t+1} is then:

$$\theta_{t+1} = \theta_t + \alpha \left(Y_t^Q - Q(S_t, A_t; \theta_t) \right) \nabla_{\theta_t} Q(S_t, A_t; \theta_t)$$

where α is a scalar step size, and the target Y_t^Q is defined as:

$$Y_t^Q \equiv R_{t+1} + \gamma \max_\alpha Q(S_{t+1}, a; \theta_t)$$

This update resembles stochastic gradient descent, updating the current value $Q(S_t, A_t; \theta_t)$ toward a target value Y_t^Q.

13.15.2   Deep Q Networks (DQN)

A deep Q network (DQN) is a multilayered neural network that for a given state s outputs a vector of action values $Q(s; \_; \theta)$, where θ are the parameters of the network. For an n-dimensional state space and an action space containing m actions, the neural network is a function from $\mathbb{R}^n$ Rn to $\mathbb{R}^m$. Two important ingredients of the DQN algorithm as proposed by [5] are the use of a target network and the use of experience replay. The target network, with parameters θ^-, is the same as the online network except that its parameters are copied every τ steps from the online network, so that $\theta_t^- = \theta_t$ and kept fixed on all other steps. The target used by DQN is then:

$$Y_t^{DQN} \equiv R_{t+1} + \gamma \max_{\alpha} Q\left(S_{t+1}, a; \theta_t^-\right)$$

For the experience replay, observed transitions are stored for some time and sampled uniformly from this memory bank to update the network. Both the target network and the experience replay dramatically improve the performance of the algorithm [5].

13.15.3   Double Deep Q-Networks (Double DQN)

The max operator in standard Q-learning and DQN, we use the same values both to select and to evaluate an action. This makes it more likely to select overestimated values, resulting in overoptimistic value estimates. To prevent this, we can decouple the selection from the evaluation. This is the idea behind Double Q-learning [6]. In the original Double Q-learning algorithm, two value functions are learned by assigning each experience randomly to update one of the two value functions, such that there are two sets of weights, θ and θ'. For each update, one set of weights is used to determine the greedy policy and the other to determine its value. For a clear comparison, we can first untangle the selection and evaluation in Q-learning and rewrite its target as:

$$Y_t^Q \equiv R_{t+1} + \gamma Q\left(S_{t+1}, \underset{\alpha}{\operatorname{argmax}}\, Q\left(S_{t+1}, a; \theta_t\right)\right)$$

The Double Q-learning TD-error can then be written as:

$$Y_t^{DoubleDQN} \equiv R_{t+1} + \gamma Q\left(S_{t+1}, \underset{\alpha}{\operatorname{argmax}}\, Q\left(S_{t+1}, a; \theta_t\right); \theta_t'\right)$$

Notice that the selection of the action, in the argmax, is still due to the online weights θ_t. This means that, as in Q-learning, we are still estimating the value of the greedy policy according to the current values, as defined by θ_t. However, we use the second set of weights θ_t' to fairly evaluate the value of this policy. This second set of weights can be updated symmetrically by switching the roles of θ and θ'.

The DDQN agent uses two identical estimator DQN networks (Figure 13.16) that are updated differently and so hold different weights. The second estimator DQN network is a copy of the main neural network from some time in the past—typically from the last episode. The loss function in the DDQN is provided as follows:

$$L(\theta_i) = E\left[\left(Y_t^{DoubleDQN} - Q(S_t, A_t; \theta_t)\right)^2\right]$$

In reference to both Double Q-learning and DQN, we refer to the resulting algorithm as Double DQN. Its update is the same as for DQN but replacing the target network with $Y_t^{DoubleDQN}$. In comparison to Double Q-learning, the weights of the second network are replaced with the weights of the target network for the evaluation of the current greedy policy. The update to the target network stays unchanged from DQN and remains a periodic copy of the online network. This version of Double DQN is perhaps the minimal possible change to DQN toward Double Q-learning. The goal is to get most of the benefit of Double Q-learning while keeping the rest of the DQN algorithm intact for a fair comparison and with minimal computational overhead.

One learns during the experience replay, just like DQN does, and the other one is a copy of the last episode of the first model. The Q-value is computed with this second model. In DQN, Q-value is calculated with the reward added to the next state maximum Q-value. Obviously, if every time the Q-value calculates a high number for a certain state, the value that is obtained from the output of the neural network for that specific state will become higher every time. Each output neuron value will get higher and higher until the difference between each output value is high. Now if let's say for state s action a is a higher value than action b, then action a will get chosen every time for state s. Now consider if for some memory experience action b becomes the better action for state s, then, since the neural network is trained in a way to give a much higher value for action a when given state s, it is difficult to train the network to learn that action b is the better action in some conditions. DQN agent uses a deep neural network to learn the Q-value function. DQN has shown itself to be a powerful algorithm for discrete action-space environments and problems and is a notable milestone in the history of deep reinforcement learning when DQN mastered Atari Games.

13.16   NEURAL FITTED Q-NETWORK (NFQ)

Neural Fitted Q-Network (NFQ) is an instance of the Fitted Q-Iteration family of algorithms, where the regression algorithm is realized by a multilayer perceptron (MLP). It consists of two major steps:

- The generation of the training set P and
- The training of these patterns within a multilayer perceptron.

A *multilayer perceptron* (MLP) is a class of feedforward artificial neural network (ANN). MLP utilizes a supervised learning technique called backpropagation for training. Its *multiple layers* and nonlinear activation distinguish MLP from a linear *perceptron*. It can distinguish data that is not linearly separable. In neural fitted Q-learning (NFQ) [7], the state can be provided as an input to the Q-network and a different output is given for each of the possible actions. This provides an efficient structure that has the advantage of obtaining the computation of $\max_{a' \in A} Q(s', a'; \theta_k)$ in a single forward pass in the neural network for a given s'. The Q-values are parameterized with a neural network $Q(s, a; \theta_k)$, where the parameters θ_k are updated by stochastic gradient descent (or a variant) by minimizing the square loss:

$$L_{DQN} = \left(Q(s, a; \theta_k) - Y_k^Q \right)^2$$

Thus, the Q-learning update results in updating the parameters:

$$\theta_{k+1} = \theta_k + \alpha \left(Y_k^Q - Q(s, a; \theta_k) \right) \nabla_{\theta_k} Q(s, a; \theta_k)$$

where α is a scalar step size called the learning rate. Note that using the square loss is not arbitrary. Indeed, it ensures that $Q(s, a; \theta_k)$ should tend without bias to the expected value of the random variable Y_k^Q (Note: The minimum of $E\left[(Z - c)^2 \right]$ occurs when the constant c equals the expected value of the random variable Z). Hence, it ensures that $Q(s,a;\theta_k)$ should tend to $Q^*(s,a)$ after many iterations in the hypothesis that the neural network is well-suited for the task and that the experience gathered in the dataset D is sufficient. When updating the weights, one also changes the target. Due to the generalization and extrapolation abilities of neural networks, this approach can build large errors at different places in the state-action space.

Note that even fitted value iteration with linear regression can diverge. However, this drawback does not happen when using linear function approximators that have only interpolation abilities such as kernel (kernel means linearization of polynomials)-based regressors (k-nearest neighbors, linear and multilinear interpolation, and others) or tree-based ensemble methods [8]. However, these methods have not proved able to handle successfully high-dimensional inputs. Therefore, the contraction mapping property of the Bellman operator in the equation:

$$\left(BK\right)\left(s,a\right) = \sum_{s'\in S} T\left(s,a,s'\right)\left(R\left(s,a,s'\right)+\gamma \max_{a'\in A} K\left(s',a'\right)\right)$$

is not enough to guarantee convergence.

It is verified experimentally that these errors may propagate with this update rule and, consequently, convergence may be slow or even unstable [7]. Another related damaging side-effect of using function approximators is the fact that Q-values tend to be overestimated due to the max operator [6]. Because of the instabilities and the risk of overestimation, specific care has been taken to ensure proper learning.

13.17   ADVANTAGE ACTOR-CRITIC NETWORK

In RL, we usually want to increase the probability of actions in a trajectory proportional to how high the return of the reward. Policy gradient algorithm is a policy iteration approach where policy is directly manipulated to reach the optimal policy that maximizes the expected return. For example, the policy gradient (PG) can be described as follows:

$$\nabla_\theta J\left(\theta\right) = \mathrm{E}_\tau\left[\sum_{t=0}^{T-1}\nabla_\theta \log \pi_\theta\left(a_t \mid s_t\right)G_t\right]$$

However, the variance of the policy gradient can be very high. The advantage function can significantly reduce variance of policy gradient. We have seen that the advantage function is defined as follows:

$$A^\pi\left(s_t,a_t\right) = Q^\pi\left(s_t,a_t\right)-V^\pi\left(s_t\right)$$

$$L_i\left(\theta_i\right) = \mathrm{E}\left[\left(r +\gamma \max_{a'} Q\left(s',a';\theta^-\right)-Q\left(s,a;\theta_i\right)\right)^2\right]$$

We refer to this method as one-step Q-learning because it updates the action value $Q(s,a)$ toward the one-step return $r +\gamma \max_{a'}\cdot Q\left(s',a';\theta^-\right)$. One drawback of using one-step methods is that obtaining a reward r only directly affects the value of the state action pair (s, a) that led to the reward. The values of other state action pairs are affected only indirectly through the updated value $Q(s,a)$. This can make the learning process slow since many updates are required to propagate a reward to the relevant preceding states and actions. One way of propagating rewards faster is by using n-step returns. In n-step Q-learning, $Q(s,a)$ is updated toward the n-step return defined as follows:

$$r_t + \gamma r_{t+1} + \ldots + \gamma^{n-1} r_{t+n-1} + \max_a \gamma^n Q\left(s_{t+n}, a\right)$$

This results in a single reward r directly affecting the values of n preceding state action pairs. This makes the process of propagating rewards to relevant state-action pairs potentially much more efficient. In contrast to Value-Based methods, Policy-Based Model-Free methods directly parameterize the policy $\pi\left(a \mid s; \theta\right)$ and update the parameters θ by performing, typically approximate, gradient ascent on $E\left[R_t\right]$. One example of such a method is the RL family of algorithms. Standard RL updates the policy parameters θ in the direction $\nabla_\theta \log \pi_\theta\left(a_t \mid s_t\right) R_t$ (note that we have used both R and G for cumulative rewards), which is an unbiased estimate of $\nabla_\theta E\left[R_t\right]$. It is possible to reduce the variance of this estimate while keeping it unbiased by subtracting a learned function of the state $b_t\left(s_t\right)$ known as a baseline [9], from the return. The resulting gradient is $\nabla_\theta \log \pi_\theta\left(a_t \mid s_t; \theta\right)\left(R_t - b_t\left(s_t\right)\right)$.

A learned estimate of the value function is commonly used as the baseline $b_t\left(s_t\right) \approx V^\pi\left(s_t\right)$ leading to a much lower variance estimate of the policy gradient. When an approximate value function is used as the baseline, the quantity $R_t - b_t\left(s_t\right)$ used to scale the policy gradient can be seen as an estimate of the advantage of action a_t in state s_t as follows:

$$A^\pi\left(s_t, a_t\right) = Q^\pi\left(s_t, a_t\right) - V^\pi\left(s_t\right)$$

It is because R_t is an estimate of $Q^\pi\left(s_t, a_t\right)$ and $b_t\left(s_t\right)$ is an estimate of $V^\pi\left(s_t\right)$. This approach can be viewed as an actor-critic architecture where the policy $\pi_\theta\left(s_k\right)$ is the actor, and the baseline b_t is the critic [4].

If choosing a_t is better than following our policy, then $A^\pi\left(s_t, a_t\right) > 0$. If it is worse than following our policy, then $A^\pi\left(s_t, a_t\right) < 0$. If a_t is exactly the action suggested by our policy, then $A^\pi\left(s_t, a_t\right) = 0$.

If we use V^π as our baseline function $b_t\left(s_t\right)$, then the policy gradient becomes:

$$\nabla_\theta J\left(\theta\right) = \mathrm{E}_\tau \left[\sum_{t=0}^{T-1} \nabla_\theta \log \pi_\theta\left(a_t \mid s_t\right) A^\pi\left(s_t, a_t\right)\right]$$

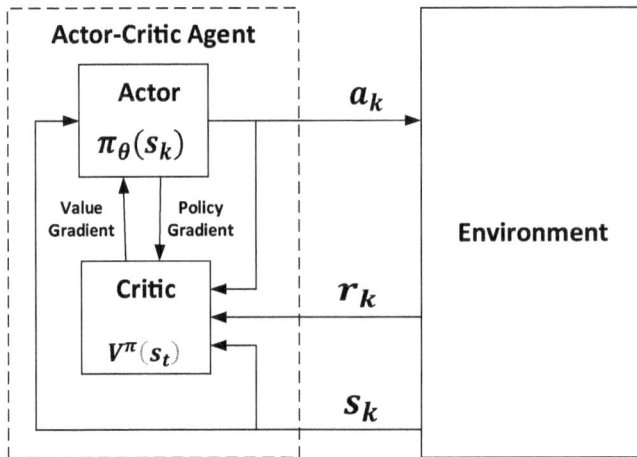

FIGURE 13.14   Actor-Critic Network.

Using the advantage is incredibly intuitive. If the advantage is positive, then we increase the log probability of the action a_t associated with that advantage. Likewise, if it's negative, then we decrease the log probability of the action a_t associated with that advantage. Like Q^π and V^π, it is extremely difficult to know the advantage exactly. Instead, we usually have to make an estimate for the advantage. There are several ways that we can do this. Note that, by our definition, a Q-network uses the neural network.

13.17.1 ESTIMATING Q^π

In deep Q-learning, we use a neural network to predict Q^π for each action given the state s_t. We fit our network using the Bellman equation to generate targets:

$$y_t = r_t + \gamma \max_{a_{t+1}} \left(s_{t+1}, a_{t+1} \right)$$

We then regress our output by minimizing the loss:

$$L(\theta) = \sum_{t=0}^{T} \left(y_t - Q(s_t, a_t) \right)^2$$

Note that instead of using the Bellman equation, we could have used the empirical returns G_t as the target:

$$L(\theta) = \sum_{t=0}^{T} \left(G_t - Q(s_t, a_t) \right)^2$$

The advantage of using the Bellman equation is that we can use it for continuing environments, whereas empirical returns G_t are available only in episodic environments. Deep Q-learning works on discrete action spaces by having one output of the neural network for each action. For continuous action spaces, this is not possible. One workaround is to simply provide both s_t and a_t into the network and produce a scalar output $Q(s_t, a_t)$.

13.17.2 ESTIMATING V^π

Just like in deep Q-learning, we can simply have a neural network V_ϕ that predicts V^π given the state. We have a Bellman equation for the value function as well, which we can use to generate targets:

$$y_t = r_t + \gamma V \left(s_{t+1} \right)$$

We then regress our output by minimizing the loss:

$$L(\theta) = \sum_{t=0}^{T} \left(y_t - V(s_t) \right)^2$$

Again, we could have just used the empirical returns if the environment is episodic:

$$L(\theta) = \sum_{t=0}^{T} \left(G_t - V(s_t) \right)^2$$

We can also generalize the policy gradient as follows:

$$\nabla_\theta J(\theta) = E_\tau \left[\sum_{t=0}^{T-1} \nabla_\theta \log \pi_\theta \left(a_t \mid s_t \right) \Psi_t \right]$$

where Ψ_t is one of them:

$$R_t = \sum_{k=t}^{T} r_k$$

$$G_t = \sum_{k=t}^{T} \gamma^{k-t} r_k$$

$$Q^\pi \left(s_t \right)$$

$$A^\pi \left(s_t, a_t \right) = Q^\pi \left(s_t, a_t \right) - V^\pi \left(s_t \right)$$

$$\text{TemporalError} : \delta_t^\pi = r_t + \gamma V^\pi \left(s_{t+1} \right) - V^\pi \left(s_t \right)$$

We can expand the Bellman equation for the value function for n steps as follows:

$$V^\pi \left(s_t \right) = E_r \left[G_t \mid s_t \right] = E_r \left[r_t + \gamma r_{t+1} + \gamma^2 r_{t+2} + \dots \mid s_t \right] = E_r \left[r_t + \gamma r_{t+1} + \gamma^2 r_{t+2} + \dots + \gamma^n V^\pi \left(s_{t+n} \right) \mid s_t \right]$$

$$= E_r \left[\sum_{k=t}^{n-1} \gamma^{k-t} r_k + \gamma^n V^\pi \left(s_{t+n} \right) \right]$$

The n-step TD-error is defined as follows:

$$\delta_t^n = \left[\sum_{k=t}^{n-1} \gamma^{k-t} r_k + \gamma^n V^\pi \left(s_{t+n} \right) \right] - V^\pi \left(s_t \right)$$

However, this is actually equivalent to a discount sum of 1-step TD-error:

$$\delta_t^n = \left(\sum_{i=0}^{n} \gamma^i \delta_{t+i} = \delta_t + \gamma \delta_{t+1} + \gamma^2 \delta_{t+2} + \dots \right)$$

$$= r_t + \left(\gamma V^\pi \left(s_{t+1} \right) - V^\pi \left(s_t \right) \right) + \gamma \left(r_{t+2} + \gamma^2 V^\pi \left(s_{t+2} \right) - V^\pi \left(s_{t+1} \right) \right)$$

$$+ \gamma^2 \left(r_{t+3} + \gamma^2 V^\pi \left(s_{t+3} \right) - V^\pi \left(s_{t+2} \right) \right) + \dots + \gamma^n \left(r_{t+n} + \gamma^n V^\pi \left(s_{t+n} \right) - V^\pi \left(s_{t+n-1} \right) \right)$$

$$= \left(r_t + \gamma r_{t+1} + \gamma^2 r_{t+2} + \gamma^3 r_{t+3} + \dots + \gamma^n V^\pi \left(s_{t+n} \right) \right) - V^\pi \left(s_t \right)$$

$$= \left(\sum_{i=t}^{n-1} \gamma^{i-t} r_i \right) + \gamma^n V^\pi \left(s_{t+n} \right) - V^\pi \left(s_t \right)$$

The value function cancels out leaving n-step TD-error.

Likewise, we can use the n-step TD-error as an advantage estimate $\hat{A}_n$. The generalized advantage estimation $\hat{A}_n^G$ is just an exponentially weighted average of these n-step estimations $\hat{A}_n^G(\gamma, \lambda)$:

$$
\begin{aligned}
\check{A}_n^G(\gamma, \lambda) &= (1-\lambda)\left[\check{A}_1(s_t) + \lambda \check{A}_2(s_{t+1}) + \lambda^2 \check{A}_3(s_{t+2}) + \ldots\right] \\
&= (1-\lambda)\left[\delta_t^\pi + \lambda\left(\delta_t^\pi + \gamma\delta_{t+1}^\pi\right) + \lambda^2\left(\delta_t^\pi + \gamma\delta_{t+1}^\pi + \gamma^2\delta_{t+2}^\pi\right) + \ldots\right] \\
&= (1-\lambda)\left[\delta_t^\pi\left(1 + \lambda + \lambda^2 + \ldots\right) + \delta_{t+1}^\pi\left(\lambda + \lambda^2 + \lambda^3 \ldots\right)\right. \\
&\quad \left. + \delta_{t+2}^\pi\left(\lambda^2 + \lambda^3 + \lambda^4 \ldots\right) + \ldots\right] \\
&= (1-\lambda)\left[\left(\delta_t^\pi \frac{1}{1-\lambda}\right) + \left(\delta_{t+1}^\pi \frac{\lambda}{1-\lambda}\right) + \left(\delta_{t+2}^\pi \frac{\lambda^2}{1-\lambda}\right) + \ldots\right] = \sum_{i=0}^{\infty}(\gamma\lambda)^i \delta_{t+i}^\pi
\end{aligned}
$$

So λ can let us interpolate between using the TD error as an advantage function and using the empirical returns minus $V^\pi(s_t)$ both of which we covered earlier as possible advantage estimates. By using a high value of λ, we use a larger sum of terms by discounting less in the future. This gives us a less biased estimate of the advantage but is more variable. Likewise, a low value of λ uses a smaller sum of terms by discounting more in the future. This gives us a more biased estimate but is less variable.

An **actor-critic** algorithm is a policy gradient algorithm that uses function estimation in place of empirical returns G_t in the policy gradient update. The **actor** is the policy $\pi(a \mid s)$ and the **critic** is usually a value function $V^\pi(s_t)$. The actor is trained using gradient ascent on the policy gradient, and the critic is trained via regression. The regression target can be either the empirical returns G_t or the Bellman target $r_t + \gamma V^\pi(s_t)$.

An **advantage actor-critic** is just an actor-critic that uses the **advantage** $A^\pi(s_t, a_t)$ instead of $V^\pi(s_t)$. Advantage can be estimated using any of the means described earlier. For our purposes, we will use GAE for our advantage estimate and fit our V_ϕ network using the empirical returns as the targets.

13.18   ASYNCHRONOUS ADVANTAGE ACTOR-CRITIC NETWORK

Asynchronous Advantage Actor-Critic (A3C) network implements parallel training where multiple workers in parallel environments independently update a global value function asynchronously having advantage, actor, and critic functions.

13.18.1   Asynchronous Advantage Actor-Critic

The algorithm, which we call asynchronous advantage actor-critic (A3C) [5], maintains a policy $\pi(a_t \mid s_t; \theta)$ and an estimate of the value function $V((s_t; \theta_v))$. Like our variant of n-step Q-learning, our variant of actor-critic also operates in the forward view and uses the same mix of n-step returns to update both the policy and the value-function. The policy and the value function are updated after every t_{max} actions or when a terminal state is reached. The update performed by the algorithm can be seen as follows:

$$
\nabla_{\theta'} \log \pi(a_t \mid s_t; \theta') A(s_t, a_t; \theta, \theta_v)
$$

where $A(s_t, a_t; \theta, \theta_v)$ is an estimate of the advantage function given by:

$$\sum_{i=0}^{k-1} \gamma^i r_{t+i} + \gamma^k V\left(s_{t+k};\theta_k\right) - V\left(s_t;\theta_v\right)$$

where k can vary from state to state and is upper bounded by t_{max}.

As with the Value-Based methods we rely on parallel actor-learners and accumulated updates for improving training stability. Note that while the parameters θ of the policy and θ_v of the value function are shown as being separate for generality, we always share some of the parameters in practice. We typically use a convolutional neural network that has one SoftMax output for the policy $\pi\left(a_t \mid s_t;\theta\right)$ and one linear output for the value function $V\left(s_t;\theta_v\right)$, with all non-output layers shared.

We also found that adding the entropy of the policy π to the objective function improved exploration by discouraging premature convergence to suboptimal deterministic policies. This technique is particularly helpful on tasks requiring hierarchical behavior. The gradient of the full objective function including the entropy regularization term with respect to the policy parameters takes the following form:

$$\nabla_{\theta'} \log \pi\left(a_t \mid s_t;\theta'\right)\left(R_t\_V\left(s_t;\theta_v\right) + \beta\nabla_{\theta'} H\left(\pi\left(s_t;\theta'\right)\right)\right)$$

where H is the entropy. The hyperparameter β controls the strength of the entropy regularization term. We used the standard non-centered **root mean squared propagation (RMSProp) update given by:

$$g = \alpha g + \left(1 - \alpha\right)\Delta\theta^2 \text{ and } \theta \leftarrow \theta - \eta \frac{\Delta\theta}{\sqrt{g + \epsilon}}$$

where all operations are performed elementwise. A comparison on a subset of Atari 2600 games showed that a variant of RMSProp where statistics g are shared across threads is considerably more robust than the other two methods.

We have presented asynchronous versions of four standard reinforcement learning algorithms and showed that they are able to train neural network controllers on a variety of domains in a stable manner. Our results show that in our proposed framework stable training of neural networks through reinforcement learning is possible with both Value-Based and Policy-Based methods, off-policy as well as on-policy methods, and in discrete as well as continuous domains. When trained on the Atari domain using 16 CPU cores, the proposed asynchronous algorithms train faster than DQN trained on an Nvidia K40 GPU, with A3C surpassing the current state-of-the-art in half the training time.

One of our main findings is that using parallel actor-learners to update a shared model had a stabilizing effect on the learning process of the three Value-Based methods we considered. While this shows that stable online Q-learning is possible without experience replay, which was used for this purpose in DQN, it does not mean that experience replay is not useful. Incorporating experience replay into the asynchronous reinforcement learning framework could substantially improve the data efficiency of these methods by reusing old data. This could in turn lead to much faster training times in domains like TORCS, where interacting with the environment is more expensive than updating the model for the architecture we used.

Combining other existing reinforcement learning methods or recent advances in deep reinforcement learning with our asynchronous framework presents many possibilities for immediate improvements to the methods we presented. While our n-step methods operate in the forward view [4] by using corrected n-step returns directly as targets, it has been more common to use the backward view to implicitly combine different returns through eligibility traces [4]. The asynchronous

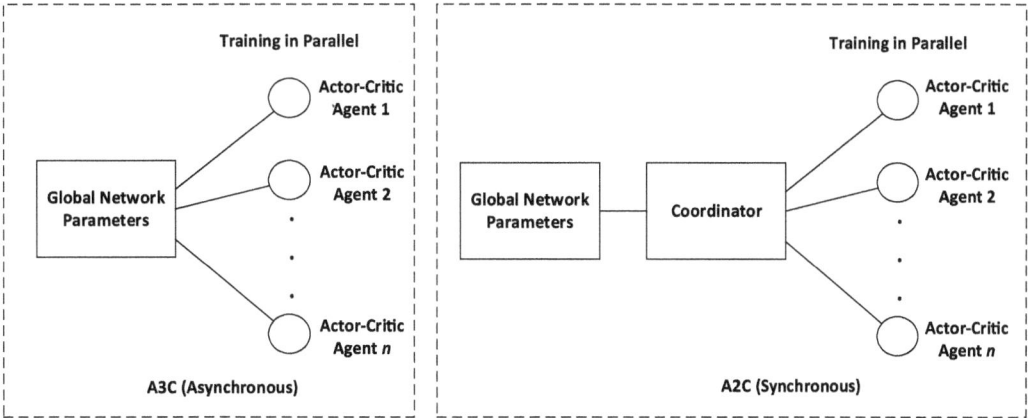

FIGURE 13.15   A3C and A2C Network.

advantage actor-critic method could be potentially improved by using other ways of estimating the advantage function, such as generalized advantage estimation. All of the Value-Based methods we investigated could benefit from different ways of reducing overestimation bias of Q-values [6]. Yet another, more speculative, direction is to try and combine the recent work on true online temporal difference methods [7] with nonlinear function approximation.

In addition to these algorithmic improvements, a number of complementary improvements to the neural network architecture are possible. The dueling architecture has been shown to produce more accurate estimates of Q-values by including separate streams for the state value and advantage in the network. The spatial SoftMax could improve both Value-Based and Policy-Based methods by making it easier for the network to represent feature coordination.

A3C is very useful when large computation power is available and the concept of transfer learning on similar environments needs to be introduced. Figure 13.15 shows the difference between A2C and A3C. A2C uses a coordinator for synchronization for all agents for accessing the global network parameters while A3C does not.

Unlike some simpler techniques which are based on either Value-Iteration methods or Policy-Gradient methods, the A3C algorithm combines the best parts of both the methods, that is, the algorithm predicts both the value function $V(s)$ as well as the optimal policy function $\pi(s)$. The learning agent uses the value of the value function (Critic) to update the optimal policy function (Actor). Note that here the policy function means the probabilistic distribution of the action space.

13.19   DUELING DEEP Q-NETWORK

Dueling Q-network has two separate streams while sharing conv parts. One stream is for state-value and the other is for advantages for each action. The Dueling architecture represents both the value and advantage functions with a single deep model whose output combines the two to produce a state-action value. The advantage function is provided as follows:

$$A^{\pi}\left(s_{t}, a_{t}\right) = Q^{\pi}\left(s_{t}, a_{t}\right) - V^{\pi}\left(s_{t}\right)$$

The value $V^{\pi}\left(s_{t}\right)$ of the advantage function measures how good it is to be in a particular state s. The Q function measures the value of choosing a particular action when in this state. The advantage function subtracts the value of the state from the Q function to obtain a relative measure of the importance of each action.

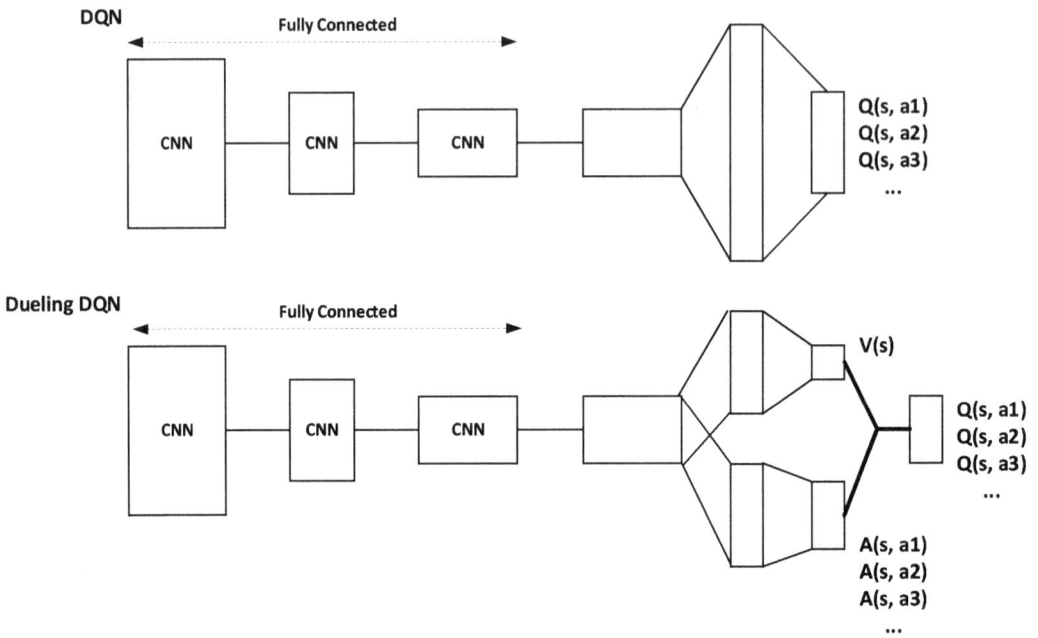

FIGURE 13.16 Dueling Network Architecture.

The key insight behind our new architecture is that for many states, it is unnecessary to estimate the value of each action choice. For example, in the Enduro game setting, knowing whether to move left or right matters only when a collision is eminent. In some states, it is of paramount importance to know which action to take, but in many other states the choice of action has no repercussion on what happens. For bootstrapping-based algorithms, however, the estimation of state values is of great importance for every state. To bring this insight to fruition, we design a single Q-network architecture, as illustrated in Figure 13.19, which we refer to as the dueling network. The lower layers of the dueling network are convolutional as in the original DQNs [5]. However, instead of following the convolutional layers with a single sequence of fully connected layers, we use two sequences (or streams) of fully connected layers. The streams are constructed such that they have the capability of providing separate estimates of the value and advantage functions. Finally, the two streams are combined to produce a single output Q function. As in [5], the output of the network is a set of Q values, one for each action.

Since the output of the dueling network is a Q function, it can be trained with the many existing algorithms, such as DDQN and State–Action–Reward–State–Action (SARSA) [1]. In addition, it can take advantage of any improvements to these algorithms, including better replay memories, better exploration policies, intrinsic motivation, and so on. The module that combines the two streams of fully connected layers to output a Q estimate requires very thoughtful design. From advantage expression, we have:

$$Q^{\pi}\left(s_{t},a_{t}\right)=A^{\pi}\left(s_{t},a_{t}\right)+V^{\pi}\left(s_{t}\right)$$

For state-value, we have:

$$V^{\pi}\left(s_{t}\right)=E_{a\sim\pi(s)}\left[Q^{\pi}\left(s_{t},a_{t}\right)\right]$$

It follows that,

$$A^{\pi}\left(s_{t}, a_{t}\right) = 0$$

For deterministic policy, $a^{*} = \arg\max_{a_{t+1} \in A} Q^{\pi}\left(s_{t}, a_{t+1}\right)$, it follows that $Q^{\pi}\left(s_{t}, a^{*}\right) = V^{\pi}\left(s_{t}\right)$ and hence $A^{\pi}\left(s_{t}, a^{*}\right) = 0$.

Note some notation changes: $t' \equiv t+1$ and any item without time notation means the present time t. Let us consider the dueling network shown in Figure 13.17, where we make one stream of fully connected layers output a scalar $V(s; \theta, \beta)$, and the other stream output an $|A|$-dimensional vector $A(s, a; \theta, \alpha)$. Here, θ denotes the parameters of the convolutional layers, while α and β are the parameters of the two streams of fully connected layers. Using the definition of advantage, we might be tempted to construct the aggregating module as follows:

$$Q(s, a; \theta, \alpha, \beta) = V(s; \theta, \beta) + A(s, a; \theta, \alpha)$$

Note that this expression applies to all (s; a) instances; that is, to express equation (7) in matrix form, we need to replicate the scalar, $V(s; \theta, \beta)$, $|A|$ times. However, we need to keep in mind that $Q(s, a; \theta, \alpha, \beta)$ is only a parameterized estimate of the true Q-function. Moreover, it would be wrong to conclude that $V(s; \theta, \beta)$ is a good estimator of the state-value function or likewise that $A(s, a; \theta, \alpha)$ provides a reasonable estimate of the advantage function.

Equation of $Q(s, a; \theta, \alpha, \beta)$ is unidentifiable in the sense that given Q we cannot recover V and A uniquely. To see this, add a constant to $V(s; \theta, \beta)$ and subtract the same constant from $A(s, a; \theta, \alpha)$. This constant cancels out resulting in the same Q value. This lack of identifiability is mirrored by poor practical performance when this equation is used directly.

To address this issue of identifiability, we can force the advantage function estimator to have zero advantage at the chosen action. That is, we let the last module of the network implement the forward mapping:

$$Q(s, a; \theta, \alpha, \beta) = V(s; \theta, \beta) + \left[A(s, a; \theta, \alpha) - \max_{a' \in |A|} A(s, a'; \theta, \alpha) \right]$$

Now, for $a^{*} = \arg\max_{a' \in A} Q(s, a'; \theta, \alpha, \beta) = \arg\max_{a' \in A} A(s, a'; \theta, \alpha, \beta)$, we obtain $Q(s, a^{*}; \theta, \alpha, \beta) = V(s; \theta, \beta)$. Hence, the stream $V(s; \theta, \beta)$ provides an estimate of the value function while the other stream produces an estimate of the advantage function. An alternative module replaces the max operator with an average:

$$Q(s, a; \theta, \alpha, \beta) = V(s; \theta, \beta) + \left[A(s, a; \theta, \alpha) - \frac{1}{|A|} \sum_{a'} A(s, a'; \theta, \alpha) \right]$$

On the one hand this loses the original semantics of V and A because they are now off-target by a constant, but on the other hand it increases the stability of the optimization: the advantages only need to change as fast as the mean, instead of having to compensate any change to the optimal action's advantage. We also experimented with a SoftMax but found it to deliver similar results to the simpler module.

Note that while subtracting the mean in $Q(s, a; \theta, \alpha, \beta)$ helps with identifiability, it does not change the relative rank of the A (and hence Q) values, preserving any greedy or ϵ- greedy policy based on Q values. When acting, it suffices to evaluate the advantage stream to make decisions.

It is important to note that $Q(s,a;\theta,\alpha,\beta)$ is viewed and implemented as part of the network and not as a separate algorithmic step. Training of the dueling architectures, as with standard Q networks (e.g., the deep Q-network [5]), requires only back-propagation. The estimates $V(s;\theta,\beta)$ and $A(s,a;\theta,\alpha)$ are computed automatically without any extra supervision or algorithmic modifications. As the dueling architecture shares the same input-output interface with standard Q networks, we can recycle all learning algorithms with Q networks (e.g., DDQN and SARSA) to train the dueling architecture.

13.20   SUMMARY

We have described the reinforcement learning in detail. RL agent and environment, optimal stationary policy, Q-Table, value of an optimal policy, policy categories, temporal difference, policy gradient, expected return, V-function, Q-function, fitted Q-learning, Deep Q-Networks, massive parallel DQN architecture, DDQN, neural fitted Q-network, advantage actor-critic network, and dueling deep Q-network have been explained. RL is one of the most dominant fields in artificial intelligence. Every RL network as described earlier needs deep attention primarily for balanced optimization without overfitting or underfitting.

REFERENCES

[1] Poole, D. L. and Mackworth, A. K., "Artificial Intelligence: Foundation of Computational Agents," Second Edition, 2017, Cambridge University Press.
[2] Jacobson, D. H. and Mayne, D. Q., "Differential Dynamic Programming," 1970, American Elsevier Publishing Company.
[3] Nair, A. et al., "Massively Parallel Methods for Deep Reinforcement Learning," 2015, arXiv:1507.04296v2 [cs.LG] 16 Jul 2015.
[4] Sutton, R. S. and Barto, A. G., "Reinforcement Learning: An Introduction," Second Edition, 1988, Francis Bach.
[5] Mnih, A. et al., "Asynchronous Methods for Deep Reinforcement Learning," 2016, arXiv:1602.01783v2 [cs.LG] 16 Jun 2016.
[6] Van-Hassett, H. et al., "Deep Reinforcement Learning with Double Q-learning," 2015, arXiv:1509.06461v3 [cs.LG] 8 Dec 2015.
[7] Riedmiller, M., "Neural Fitted Q Iteration—First Experiences with a Data Efficient Neural Reinforcement Learning Method," 2005, ECML.
[8] Ernst, D. et al., "Tree-Based Batch Mode Reinforcement Learning," Journal of Machine Learning Research. 6 (2005): 503–556, 2005.
[9] Williams, R. J., "Simple Statistical Gradient-Following Algorithms for Connectionist Reinforcement Learning," Machine Learning, 1992.

14 Artificial Intelligence Application and Network Protocol Architecture Model

The 21st century AI-enabled/standard-based communications networks have to deal with many diverse applications. AI-enable means the existing legacy applications/protocols that are enabled using AI may be placed in the backend server while the AI-standard-based applications/protocols are the new efficient standardized applications and networking protocols that do not use any legacy applications or protocols. For example, 4G/5G networks can use AI-enabled applications protocols where legacy applications protocols are processed by an AI backend server to make them more efficient. However, the emerging 6G network will inherently be new AI-standard-based communications network, where all applications and networking protocols will be designed using open standard-based AI protocols. The AI-enabled/standard-based communications network protocol architecture is depicted in Figure 14.1. It is using the same OSI architecture model, but all protocol layers are AI-enabled or AI-standard-based.

The present legacy applications in each layer is AI-enabled to make them more efficient. When AI standard-based protocols are standardized by open standards organizations, those new AI standard-based protocols will also be used in each layer. Note that AI-aware applications services could be both AI-enabled or AI standard-based. For simplicity, we will use the term AI applications that may indicate AI-enabled or AI-standard-based applications or protocols.

The 21st-century communications networks must deal with many diverse applications with the span of space-tier, airborne-tier, unmanned airborne vehicle (UAV)-tier, and ground (manned and unmanned)-tier along with mobile cellular wireless networks, mobile ad hoc networks (MANETs), and/or fixed wireline networks. Figure 14.2 shows a proposed high-level view of the multi-domain communications network architecture [1].

Communications networks operations are also becoming increasingly diverse in their nature. Each of the applications like voice/multimedia-over-Internet protocol like session initiation protocol (SIP), messaging, instant messaging, situational awareness (SA), command & control, embedded training, forward observer training, live cyberwarfare simulation, and others need to have only one integrated picture containing the history, current status, and potential consequences of actions especially in the cyberwarfare environment. So each of these applications need massive information processing power in the shortest possible time.

The condition of information overload occurs when one is unable to process the information presented into coherent SA, for example. With the rapidly expanding ability to collect data in real-time or near-real-time about many locations and providing data abstractions to the communications at different levels from the command-and-control operational center to individual field personnel, the danger of information overload has grown significantly.

A network operator or user may need to understand the global situational awareness and how the various teams are expected to move through an environment, whereas a single user may be concerned only with a very limited area. Similarly, a medic may need health records and a route to an injured person, whereas a forward observer may need a few days' worth of reconnaissance information to detect unusual or unexpected adversaries' or attackers' actions. Ideally, the AI/ML/DL applications would be aware of these various tasks such as the mission plans (including any contingencies) and the current roles that any user may fulfill at a given time. These users' requirements and capabilities have specifically driven the development of AI/ML/DL applications [2].

DOI: 10.1201/9781003499466-14

AI-aware Applications Services
Cybersecurity, O&M, NLP, Speech, Analytics/Prediction/Forecast, Expert System, Robotics, Vision, Mood/Sentiment Analysis, and others

AI-enabled/standard-based Applications
NLP Platforms (Context Extraction/Classifications/Machine Transformation/ Questions & Answers/Test Correlation), Expert System Platform, Knowledge Base (KB), Ontology, Vision Platform(Image Processing/Machine Vision), Speech Platform (Text-to-Speech [TTS]/Speech-to-Text [STT]), Robotics Platform and others

AI-enabled/standard-based Middleware

AI-Standard-based Middleware
ML/DL Algorithms (Linear/Logistics Regression, Linear Discrimination Analysis, Gradient Boosted Decision Tree (XG Boost), Factorization Machines, K-means/ Extraction Maximizing Clustering, Random Forest, Naïve Bayes, Principal Component Analysis (PCA), Linear Learner, Support Vector Machine (SVM), Belief Network, Hidden Markov Model (HMM), Profile Hidden Markov Model (PHMM), Fuzzy Systems, Long-Term Short-Term Memory (LSTM), Recurrent Neural Network (RNN), Convolution Neural Network (CNN), Neural Turing Machine (NTM), ML/DL Tools (OpenAI, TensorFlow, Keras, H2O, PyToch, Caffe, Gluon), and others

AI-enabled Middleware
All legacy middleware that are made more efficient using AI

AI-enabled/standard-based Transport Protocols
AI-enabled Legacy Transport Protocols (TCP, UDP, TLS-TCP, DTLS-UDP, HTTP, HTTPS)/AI standard-based New Transport Protocols

AI-enabled/standard-based Network and Routing Protocols
AI-enabled Legacy Network and Routing Protocols (IPv4, IPv6, Routing Protocol [e.g., OSPFv3, OSPFv3-MDR, OSPFv3-CDR, DSR, OLSR, BGP])/New AI standard-based New Network and Transport Protocols

AI-enabled/standard-based Link/Medium Access Control (MAC) Protocols
AI-enabled Legacy Link/MAC Protocols (CSMA/CA, CSMA/CD, TDMA, CDMA, FDMA, Hybrid (TDMA/CDMA/FDMA), OFDMA)/New AI standard-based New Link/MAC Protocols

AI-enabled/standard-based Physical (PHY) Layer
AI-enabled 4G/5G, AI-based 6G, AI-enabled/AI standard-based Fiber/Optical/ Dense Wavelength Division Multiplexing (DWDM)

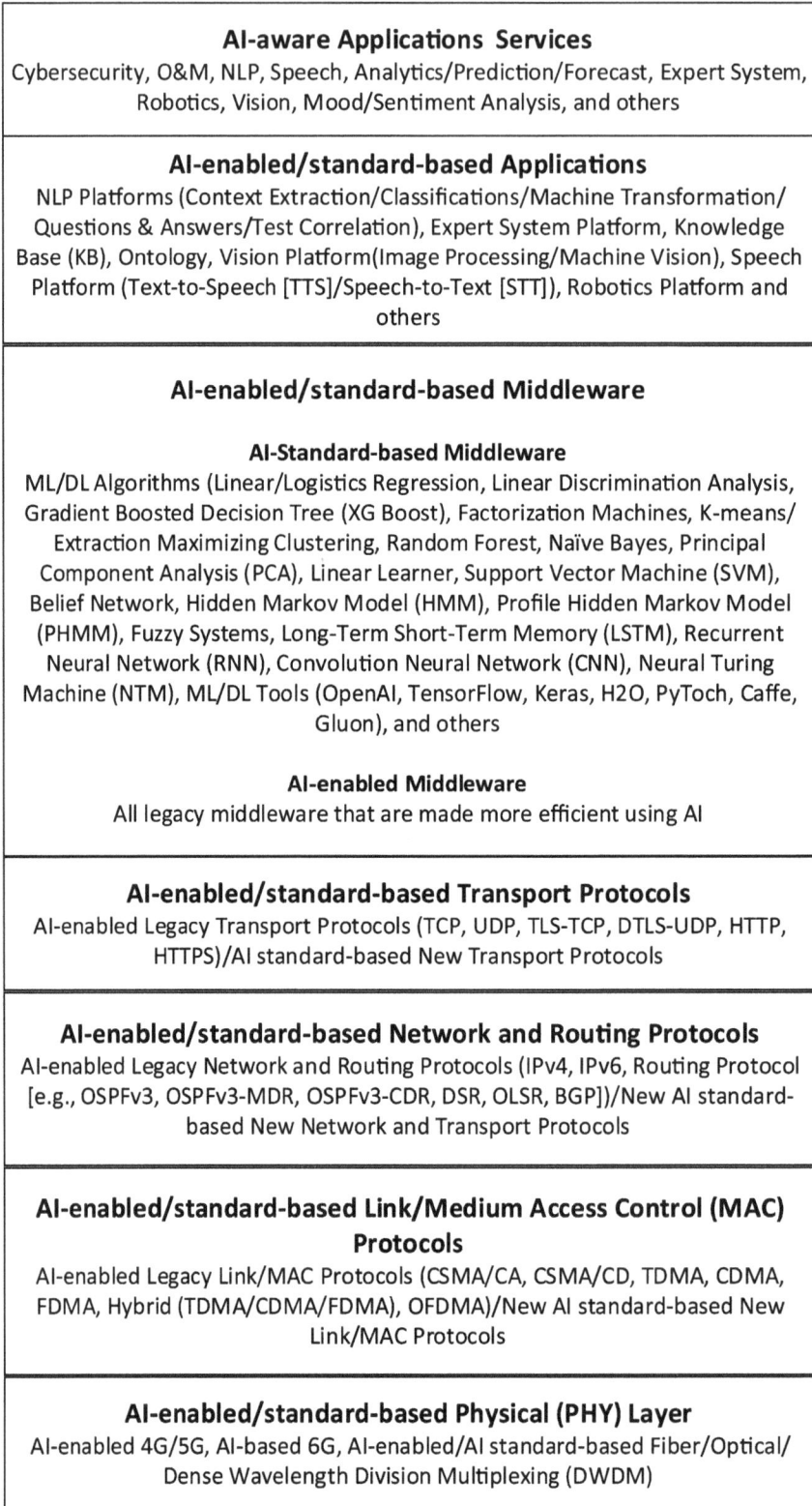

FIGURE 14.1 AI Communications Network Protocol Architecture.

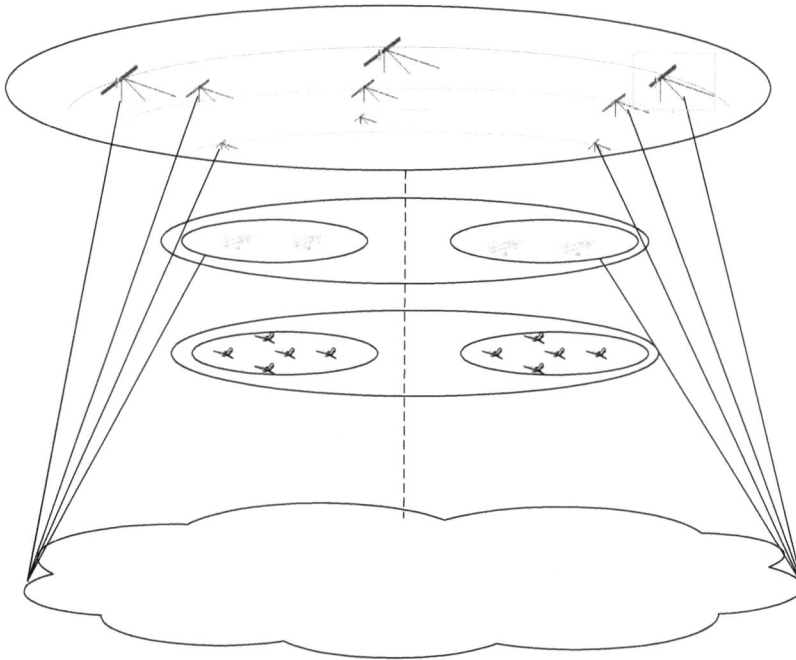

FIGURE 14.2 High-Level View of Multi-Domain Communications Network Architecture.

It should also be evident at this point that the various AI/ML/DL applications bridge two somewhat disparate fields: computation and communication. For example, SA compels that the visual representations of data need to be introduced. Similarly, all other applications like sensors, robots, vehicles, computers, phones, servers, and other communications entities will be using AI/ML/DL applications. The high-level data abstractions that satisfy the requirements of all diverse applications in generic application-agnostic ways are also an active research area.

AI needs to integrate a diverse amount of information from different applications including the outputs of ML/DL algorithms. Consequently, it can limit the types of abstractions that make sense for a given application and push the application designer to create new methods of understanding perceptual or cognitive cues that go beyond typical human sensory experiences. The high-level abstractions, which are independent of programming languages used and supported dataset formats, facilitate for development of application programing interfaces (APIs) to draw on individual algorithms from different products or to compose multiple algorithms to solve complex tasks using AI/ML/DL.

14.1 ARTIFICIAL INTELLIGENCE-AWARE APPLICATIONS SERVICES

The AI-aware application services may consist of cybersecurity, O&M, NLP, speech, analytics/prediction/forecast, expert system, robotics, vision, mood/sentiment analysis, and others. These services are created based on AI-enabled/standard-based applications of the lower AI-enabled/standard-based middleware layer.

14.2 ARTIFICIAL INTELLIGENCE-ENABLED/STANDARD-BASED APPLICATIONS

This layer consists of AI applications that are usually used for building AI application services as follows:

- NLP Platforms: Context Extraction/Classifications/Machine Transformation/Questions & Answers/Test Correlation
- Expert System Platform
- Knowledge Base (KB)
- Ontology
- Vision Platform: Image Processing/Machine Vision
- Speech Platform: Text-to-Speech (TTS) and Speech-to-Text (STT)
- Robotics Platform
- Others

These AI applications are critical for performing all the works that we see in any AI systems, products (e.g., ChatGPT, BARD, and others), and services. In turn, these AI services invoke ML/DL algorithms that we have termed as the common ML/DL middleware infrastructure for performing computations on behalf of applications. As described earlier, one or more ML/DL algorithms could be invoked by AI applications to find the best performances for satisfying mission objectives.

14.3   ARTIFICIAL INTELLIGENCE-ENABLED/STANDARD-BASED MIDDLEWARE INFRASTRUCTURE

We have divided AI-enabled/standard-based Middleware into two parts: AI Standard-based Middleware standardized by open or public standards organization and AI-enabled Middleware where legacy existing middleware is enabled using AI by proprietary means.

14.3.1   ARTIFICIAL INTELLIGENCE-STANDARD-BASED MIDDLEWARE

The most important aspect of this layer is the AI standard-based middleware that is being used for development of AI-enabled and AI standard-based applications. Also, we use the term "AI Middleware Infrastructure" for this. This middleware is a set of AI/ML/DL algorithms. A list of these AI standard-based middleware (algorithm) and tools that are expected to be standardized by the open or public standards organizations is provided here:

- Linear/Logistics Regression
- Linear Discrimination Analysis
- Gradient Boosted Decision Tree (XG Boost)
- Factorization Machines
- K-means/Extraction Maximizing Clustering
- Random Forest
- Naïve Bayes
- Principal Component Analysis (PCA)
- Independent Correlation Analysis (ICA)
- Linear Learner
- Support Vector Machine (SVM)
- Belief Network
- Hidden Markov Model (HMM)
- Profile Hidden Markov Model (PHMM)
- Fuzzy Systems
- Q-Learning Learning Model (Reinforcement Learning)
- Long-Term Short-Term Memory (LSTM)
- Recurrent Neural Network (RNN)
- Convolution Neural Network (CNN)
- Neural Turing Machine (NTM)

- Genetics Algorithm (GA)
- ML/DL Tools
  - OpenAI
  - TensorFlow
  - Keras
  - H2O
  - PyToch
  - Caffe
  - Gluon
  - others

14.3.2 ARTIFICIAL INTELLIGENCE-ENABLED MIDDLEWARE

Artificial Intelligence-Enabled Middleware is the existing legacy middleware that is AI-enabled using proprietary means. As indicated earlier, the existing legacy middleware can be enabled using AI server in the backend.

14.4 ARTIFICIAL INTELLIGENCE-ENABLED/STANDARD-BASED TRANSPORT PROTOCOLS

Legacy transport protocols that are used in today's networks are described as follows as examples: Transmission Control Protocol (TCP), User Datagram Protocol (UDP), Transport Layer Security (TLS) with TCP (TLS-TCP), Datagram Transport Layer Security (DTLS) with UDP (DTLS-UDP). In addition, Hypertext Transfer Protocol (HTTP) and Hypertext Transfer Protocol Secure (HTTPS) are used on the top of transport protocol to transfer information between networked devices especially for webservices.

TCP has flow control mechanisms, and it becomes harder to design the optimal feedback-action mapping as networks become more complex and dynamic. AI can be applied to this legacy TCP flow control using for example federated learning, reinforcement, supervised, and other machine leaning for improving performances with minimal packet losses, improved delay, and higher throughput capacity.

UDP is an unreliable protocol that drops packets over the network whilst at the same time crucial contents of information tend to be lost along the way, hence degrading overall performance. Voice, video, and data are sent over the UDP. Performances of these applications sent over the UDP are not guaranteed. Enabling UDP using AI algorithms (e.g., federated learning, reinforcement learning, supervised learning), the performances (e.g., packet loss and reliability) of the unstructured data payload can be improved.

TLS security protocol is transferred over the connection-oriented TCP transport protocol. Similar to TCP, TLS-TCP can also be AI-enabled with improved performances. DTLS-UDP can also be AI-enabled similar to the AI-enabled UDP. The application layer HTTP/HTTPS protocol is usually run over the TCP transport protocol. Although an AI website builder uses machine learning to automatically generate a web page based on user inputs, performances of HTTP/HTTPS can further be improved similar to TCP controlling the transmission of the HTTP/HTTPS protocol, but researchers are yet to work on it.

14.5 ARTIFICIAL INTELLIGENCE-ENABLED/STANDARD-BASED NETWORK AND ROUTING PROTOCOLS

Legacy Network Protocol (e.g., IPv6) and Routing Protocols (Routing Protocol [e.g., Open Short Path First Version-3 (OSPFv3), OSPFv3-MANET Designated Routers (MDR), Dynamic Source Routing (DSR), Optimized Link State Routing (OLSR), Border Gateway Protocol (BGP)]) used in

5G network can be AI-enabled while emerging 6G networks will be based on new AI standard-based New Network and Routing Protocols. AI-enabled IPv6 can offer powerful programming capabilities with network path, service, and forwarding behavior meeting numerous requirements making it ideal for service-driven networks having simplicity of operations. Routing protocols require real-time link traffic information, accurate real-time data analytics and traffic prediction methods, and consistent traffic prediction over multiple datasets in modern dynamic networks. AI-based routing protocols (e.g., OSPFv3, OSPFv3-MDR, OSPFv3-CDR, DSR, and OLSR) can take care of these problems in dynamic networks having superior performances (e.g., low packet losses, and extremely low latency) with consistent traffic prediction over multiple datasets in modern networks with dynamic topologies.

BGP is a path vector protocol used to carry routing information between autonomous systems (AS). The term path vector comes from the fact that BGP routing information carries a sequence of AS numbers that identify the path of AS that a network prefix has traversed. The path information associated with the prefix is used to enable loop prevention. BGP needs to send the pre-provisioned routing information between its peers using TCP, thereby simplifying the complexity associated with designing reliability into the protocol itself. Although BGP does not use dynamic routing protocols per se, even then AI might help even for the provision-based routing among its peers. In addition, AI can enhance energy optimization, dynamic fault recovery and analysis, resource management and scheduling for both wireline and wireless networks and cell-sectorization in particular for cellular network.

14.6   ARTIFICIAL INTELLIGENCE-ENABLED/STANDARD-BASED LINK/MEDIUM ACCESS CONTROL (MAC) PROTOCOLS

Some examples of Link/Medium Access Control (MAC) protocols are listed here:

- Carrier Sense Multiple Access/Collision Avoidance (CSMA/CA)
- CSMA/Collision Detection (CD)
- Time Division Multiple Access (TDMA)
- Code Division Multiple Access (CDMA)
- Frequency Division Multiple Access (FDMA)
- Hybrid (TDMA/CDMA/FDMA)
- Orthogonal Frequency-Division Multiple Access (OFDMA)

The link/medium access control (MAC) layer channel estimation and prediction, symbol detection, and channel coding require complex mathematical models. These mathematical models provide simple close approximated mathematical formulations and cannot capture the actual characteristics of today's high-speed (e.g., millimeter-wave and terahertz) networks especially for wireless networking environments. Instead of exact models, system designers have used heuristic models in designing the transceivers. However, AI-enabled models and algorithms for link/MAC protocol overcome these challenges of legacy models and algorithms. In the data link layer, the use of deep learning for tasks such as spectrum allocation, traffic prediction, and link evaluation will offer superior result. In the same token, medium access technology that combines the two traditional MAC technologies CSMA/CA, CSMA/CD, TDMA, CDMA, FDMA, Hybrid (TDMA/CDMA/FDMA), and OFDMA can use AI including Q-learning for better performances. A good feature of reinforcement leaning is that the agent will be able to switch between the multiple modes according to the network state and thereby keeping the network performance as high as possible for all time. Similar is the case for other technologies for all other layers.

14.7 ARTIFICIAL INTELLIGENCE-ENABLED/STANDARD-BASED PHYSICAL (PHY) LAYER

The physical (PHY) layer is critical for AI because this layer provides all computing processing power for AI/ML/DL. 5G is AI-enabled as it has been standardized to have its applications AI-enabled. 4G design has no standardized proposal for its applications, but still its applications can be made AI-enabled. However, emerging new 6G technologies will be AI standard-based by design. Some of the PHY layer tasks can be physical coding, modulation classification, interference alignment, and jamming resistance. All of these PHY layer tasks can be AI-enabled at least for 4G/5G while 6G technologies will be AI standard-based.

Free Space Optics Systems (FSO) is one of the most effective solutions, especially for atmospheric turbulence due to the weather and environment structure. Free space optics system suffers from various limitations. A well-known disadvantage of FSO is its sensitivity on local weather conditions, primarily to haze and rain, resulting in substantial loss of optical signal power over the communication path. The main objective of this chapter is to evaluate the quality of data transmission using Wavelength Division Multiplexing (WDM) while highlighting several factors that will affect the quality of data transmission. The results of these analyses are to develop a system of quality-free space optics for a high data rate transmission. From the result analysis, FSO wavelength with 1,550 nm produces less effect in atmospheric attenuation. Short link range between the transmitter and receiver can optimize the FSO system transmission parameters or components. Based on the analysis, it is recommended to develop an FSO system of 2.5 Gbps with 1,550 nm wavelength and link range up to 150 km at the clear weather condition of bit-error-rate (BER)10^{-9}.

14.8 SUMMARY

In this section, we have proposed an Artificial Intelligence Protocol Architecture Model in conformity with the OSI model. This model is going to be the most important one that will allow us to build the standard-based protocol architecture providing interoperability in multi-vendor communication environments. In view of the lack of AI standard-based protocol architecture as no standard bodies have yet to standardize, we have used dual approaches: AI-enabled protocol architecture and AI-standard-based architecture. AI-enabled protocol architecture that is built with proprietary schemes will be interoperable only via open application programming interfaces (APIs) developed by each company or organization until the time AI-standard-based protocols are available. Almost all vendors have defined their own open APIs, but none of these is based on standard-based APIs developed by open standards forums/bodies. Based on the ongoing research and development publications on open standards-based protocols, we have proposed AI-standard-based protocol architecture. It will pave the way for building the open AI-standards-based protocol architecture by the standard bodies/fora such as ITU-T, IETF, and others. In the first stage, all AI application protocols can use all the AI protocols proposed in research and development publications for most efficient operations proving superior performances with the use of simple protocols in all OSI layers.

REFERENCES

[1] Roy, R. R. et al., "Commanding in Multi-Domain Formations: Vision 2050 Warfighter Cyber-Security, Command and Control Architecture," MAD Scientist Conference, 2017.
[2] Roy, R. R. et al., "Artificial Intelligence, Machine Learning/Deep Learning-based Warfighter Cybersecurity Architecture Framework," MILCOM, 2019.

15 AI-Enabled Network

15.1 AI-ENABLED PHYSICAL LAYER

The physical layer is responsible for transmission and reception of raw bits over a communication link or channel. The physical layer link/channel can be physical wire or wireless. The traffic multiplexing, modulation scheme, data rate, channel access method, physical connectors, simplex, full/half-duplex, timing, frequency, voltage level, and others are the functions of physical layer. The physical layer link/channel can be physical wire or wireless. In physical layer, modulation/demodulation is the primary function. Modulation/demodulation (Modem) is the manipulation of the amplitude, frequency, or phase of an electromagnetic (EM) wave with the intent of transmitting information. The transmitter and receiver will agree to particular modulation scheme ahead of time to allow this information sharing to occur. We are showing the basic functions of a modem that is not AI/ML/DL-enabled in Figure 15.1.

Interestingly, if a modem is AI/ML/DL-enabled as depicted in Figure 15.2 using the processes (signal acquisition, preprocessing, feature extraction, feature classification, and prediction/decision), its basic functions will still remain the same as shown in Figure 15.1, but its performances such as bit-error-rate (BER) will be almost negligible to normal, and the channel capacity will be almost nearer to normal Shannon Channel Capacity.

It has been seen that the performance of neural networks for the modulation classification task can achieve very high levels of accuracy over a large range of modulation types under nominal channel distortions. The convolutional neural network and residual neural network achieved accuracies of over 80–90% especially with the knowledge of physics of wireless channel [1]. Note that the processes of AI/ML/DL remain the same, but training with specific signal/data using specific chosen from the common set of software repositories of algorithms tuning their parameters to those specific signal/data is needed for Physical Layer (OSI Layer 1). Figure 15.3 shows a simplified view of Figure 15.2.

Note that the simplified representation of AI/ML/DL-based modem over the communication channel is shown in Figure 15.3 as an example. We can now argue that a physical modem device might be designed with specific AI/ML/DL-embedded hardware as a standalone hardware. The other alternative would be to have the common AI/ML/DL computing processes for all OSI layers in devices like smartphones, other waveforms of radios, and application servers (Figure 15.4).

The physical layer network consists of fiber/photonic/optical, dense wavelength, and other communications networking technologies that are used in the backbone network, while 4G/5G/6G

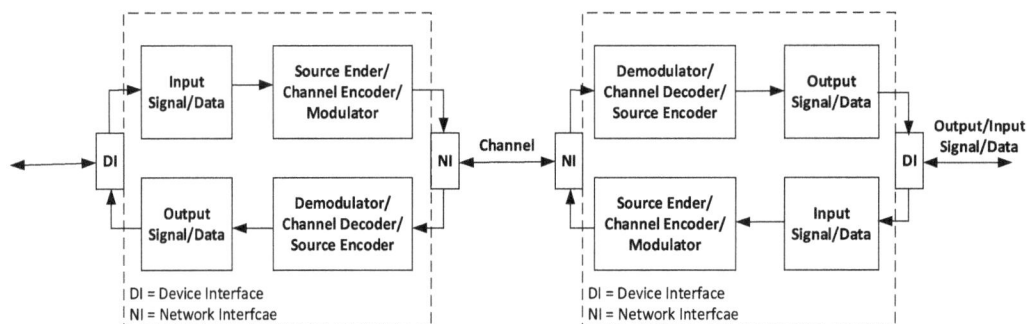

FIGURE 15.1 Basic Functions of a Modem (Modulator/Demodulator).

DOI: 10.1201/9781003499466-15

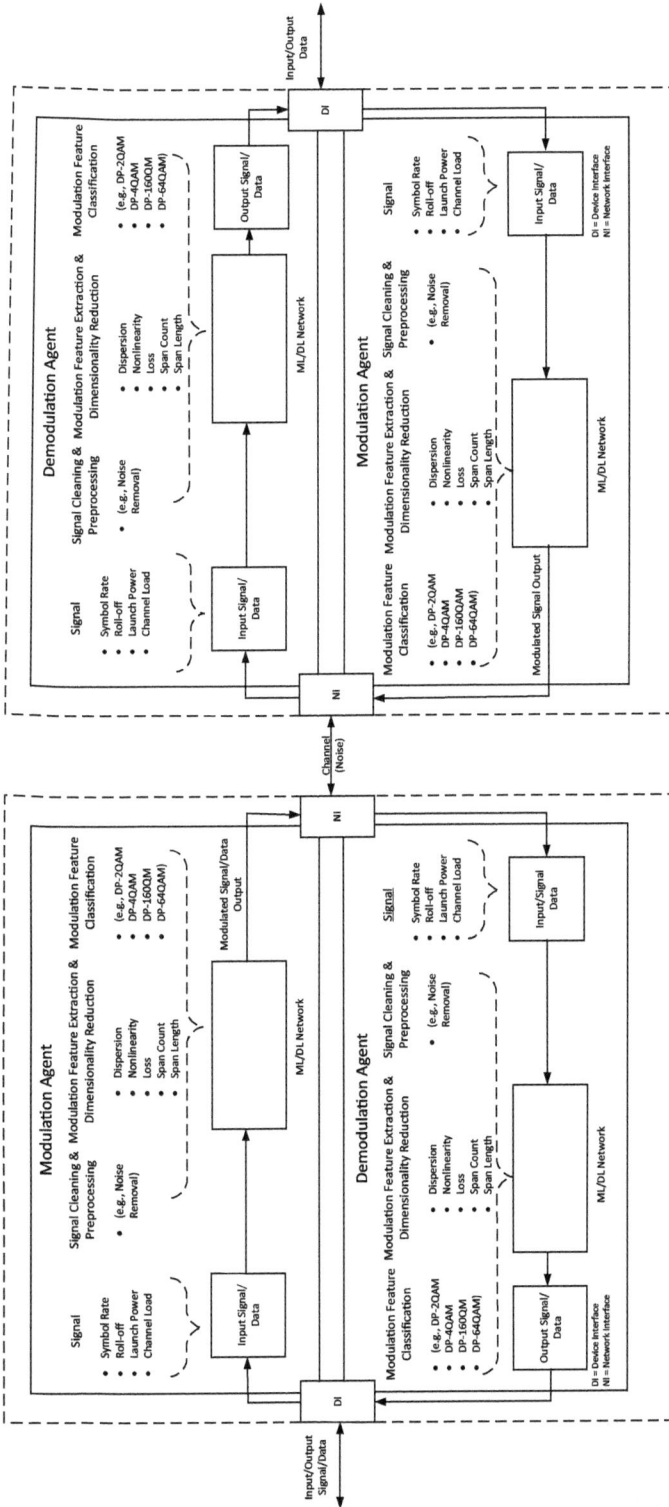

FIGURE 15.2 AI/ML/DL-Enabled Modem (Modulator/Demodulator).

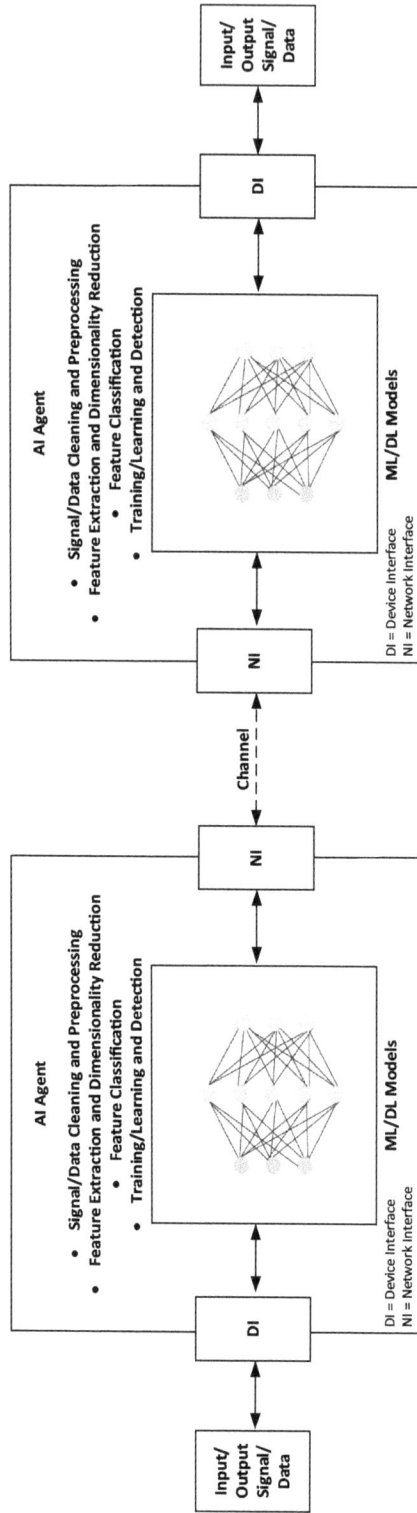

FIGURE 15.3 Simplified AI/ML/DL-Enabled Modem (Modulator/Demodulator)—Schematic Diagram.

**AI-enabled
Legacy Client**

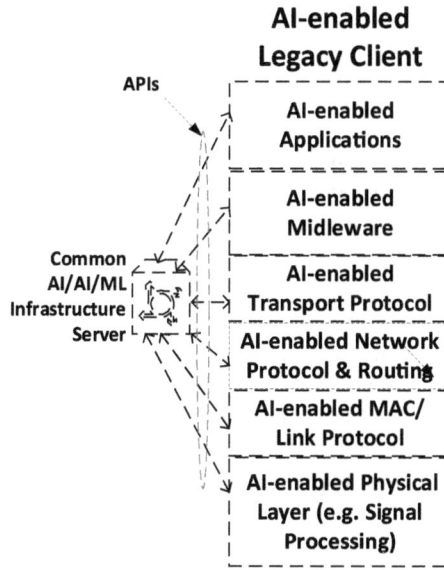

FIGURE 15.4   Common AI/ML/DL Server for All OSI Layers.

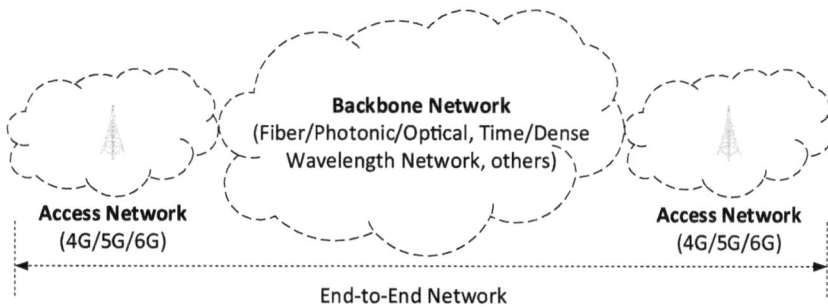

FIGURE 15.5   High-level schematic view of the End-to-End network.

wireless technologies are used in the access network connected to the backbone network. The term "generation" in the context of wireless communication typically describes to a shift in the essential characteristics of the assistance being offered, such as transmission technologies, bit rates, frequency bands, channel frequency bandwidth, and data transfer capacity. One of the most active technological fields that is expanding too quickly is the wireless age [2]. The dramatic increase in bandwidth of 6G networks will force an increase in the capacity of the backbone network. Figure 15.5 depicts a high-level schematic view of the access and backbone network.

There is a need to implement such technologies that can be integrated and adjusted to produce a more advanced and unified system because there are numerous superior technologies available. In this work, we provide a summary of several wireless telecommunications technologies, in particular, 4G, 5G, and 6G networks and a detailed comparison between them. Higher data rates and expanded capacity were supplied by third-generation (3G) mobile technology. The third generation (3G) of mobile wireless technology has been replaced by the fourth generation (4G), which offers more improved data. The Wireless World Wide Web will be made possible by 5G while 6G will combine 5G and satellite networks to offer global coverage.

15.2   AI-ENABLED LINK/MEDIUM ACCESS CONTROL LAYER

We have observed that time division multiple access (TDMA), code division multiple access (CDMA), frequency division multiple access (FDMA), orthogonal frequency-division multiple access (OFDMA), carrier sense multiple access with collision avoidance (CSMA/CA), Time-Shared Full Duplex (TSFD), and pure ALOHA, slotted ALOHA, and other wireless access protocols are used in the link/medium access layer (MAC). On the other hand, the data link control is responsible for reliable transmission of message over transmission channel by using techniques like framing, error control, and flow control.

The key is that the MAC protocol regulates the usage of media (e.g., audio, video, and/or data), and it is done through a channel access mechanism. The channel access mechanism divides the main resource through multiplexing method allowing several data streams or signals to share the same communication channel or physical medium between communicating nodes, the radio channel, by regulating the use of it. The MAC protocol tells each node when it can transmit and when it is expected to receive data. The channel access mechanism is the core of the MAC protocol. Moreover, MAC implements acknowledgment schemes and retransmissions in case of errors over the channel/time slot.

With the use of AI/ML/DL, the implementations of all those link/MAC protocols can be simplified and errors and losses over wireless networks can be minimized and hence the overall response time over the network is improved. [3–5] The medium access control (MAC) protocol in wireline/wireless networks with multiple levels of heterogeneity. The current best practice for programming MAC logic in an embedded wireless node is to implement known protocols such as ALOHA, Slotted-ALOHA, carrier sense multiple access (CSMA), and CSMA with collision avoidance (CSMA-CA) (i.e., Wi-Fi, BT, etc.) depending on the available lower layer hardware support. The choice of such protocols is often driven by heuristics and experience of network designers. While such methods provide a standard method for network and protocol deployment, they do not necessarily maximize the MAC layer performance in a fair manner, especially in the presence of network and data heterogeneity. An example is when nodes without carrier-sensing abilities run AOLHA family of protocols, their performance starts degrading due to collisions when the application traffic load in the network exceeds an optimal level. Such problems are further compounded in the presence of various forms of heterogeneity in terms of Acknowledgement (ACK), No-ACK, Fragmentation/Aggregation, Fragmentation Threshold, Back-off, Request-To-Send (RTS)/Clear-To-Send (CTS), Network Allocation Vector (NAV), NAV Timeout, Contention Window (CW), Traffic Load, Topology, Node-specific Access priorities. The key reason for such performance gap is that the nodes are statically programmed with a protocol logic that is not cognizant of time-varying load situations and such heterogeneities. The AI/ML/DL-Enabled MAC allows wireless nodes to learn how to detect such conditions and change transmission strategies in real time for maximizing the network performance, even under node-specific access prioritization.

The key concept of AI/ML/DL-Enabled MAC is to model the MAC layer logic as, for example, a Markov Decision Process (MDP) and solve it dynamically using, for example, Reinforcement Learning (RL) as a temporal difference solution under varying traffic and network conditions. An MDP solution requires the correct set of transmission actions taken by the network nodes, which act as the MDP agents. RL provides opportunities for the nodes to learn on the fly without the need of any prior training data. The example model allows to provision for heterogeneous traffic load, network topology, and node-specific priority while striking the right balance between node-level and network-level performance. Learning adjustments to temporal variations of such heterogeneities and access priorities are also supported by leveraging the inherent real-time adaptability of reinforcement learning. It is shown that the nodes can learn to self-regulate collisions to attain theoretically maximum MAC-level performance at optimal loading conditions. For higher load, the nodes learn to adjust their transmit probabilities to reduce the effective load, maintain low levels of collisions, thus maintaining the maximum MAC level performance.

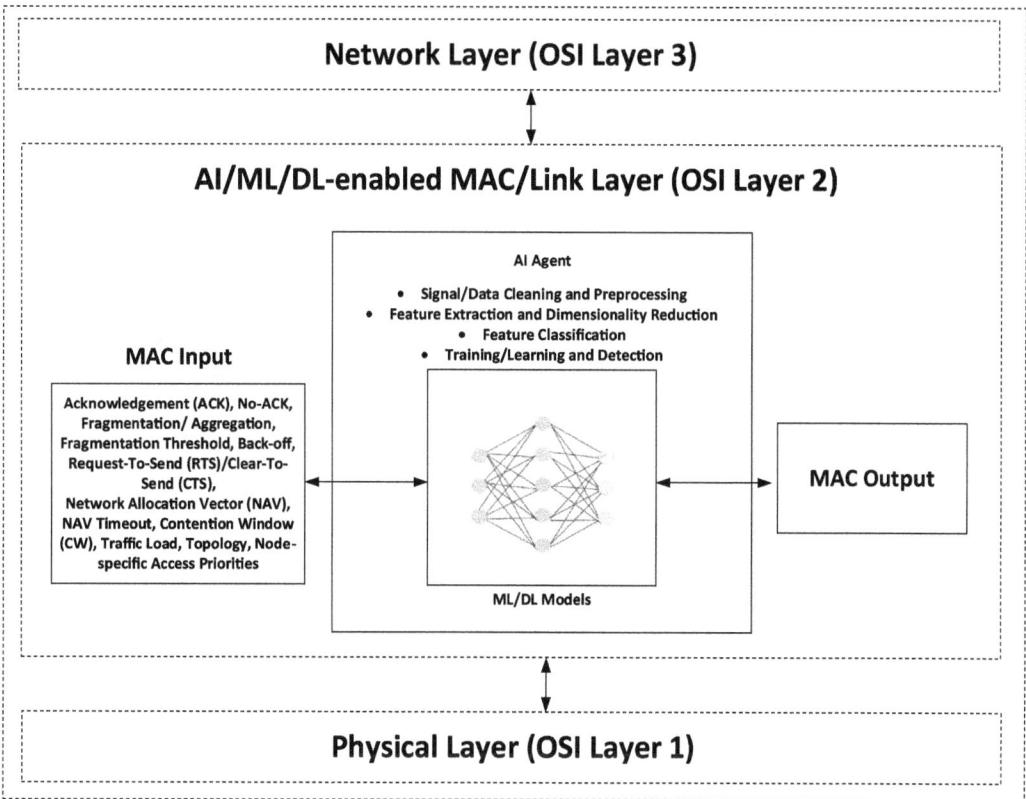

FIGURE 15.6 Simplified AI/ML/DL-Enabled Medium Access Protocol (MAC) Schematic Diagram.

The MAC/Link layer being in OSI Layer 2 works between the network layer (OSI Layer 3) and the physical layer (OSI Layer 1), and the traffic payload may come from both layers 1 and 3 depending on the direction of transmission. However, the general mechanism of AI/ML/DL-enabled MAC remains the same as expected in the legacy non-AI/ML/DL-enabled network. The description of the AI/ML/DL-enabled MAC layer is depicted in Figure 15.6.

The specific characteristics of AI/ML/DL-enabled MAC are as follows: first, a Markov Decision Process (MDP) can be used as the MAC layer generic logic solving the problems using reinforcement learning (RL). RL can make the agents/nodes self-regulate their individual traffic loads to attain and maintain the theoretically maximum throughput even when the application-level traffic load is increased beyond the point of optimality. Third, the RL mechanism can be enhanced to handle heterogeneous traffic load. Fourth, node-level access priorities can be incorporated in the native learning process. Fourth, the core concept of MDP formulation and RL solution can be used for non-fully connected network topologies. Finally, the wireline/wireless nodes with AI/ML/DL-enabled MAC can use more simplifies run a rudimentary MAC logic without relying on carrier sensing for example and other complex features from the underlying hardware. In other words, the developed mechanisms could even be more suitable for very simple transceivers that are found in low-cost wireless sensors, Internet-of-Things (IoTs), and other embedded devices.

Artificial intelligence (AI) enables machines to mimic human brain-like intelligence and is used to optimize beyond 5G (B5G) MAC. The capabilities of AI include natural language processing, knowledge-based decisions, and perception. ML is the subset of AI. ML is the general technique of AI that can learn directly from structured and unstructured data provided by information technology without any explicit programming. The ML techniques that can learn from labeled and unlabeled datasets for prediction are termed supervised learning.

The present research papers that are available are directed to devices and methods that provide a user of a decentralized asynchronous parallel-configured wireless communication system for voice, data, and live digital video streaming communication with the ability to select various communication paths and calling bandwidths as needed. The system uses a novel Time-Shared Full Duplex (TSFD) protocol for communications. The TSFD protocol allows the transmission of live digital video signals from one wireless device to another wireless device. The system provides local communication as well as optional links to external networks and does not require a synchronous centralized switching center. It further provides secure operation, emergency notification, and a way to collect revenue from the system and allows for control of the operational state of the internal network and optional remote control of the operational state control of systems external to the network.

Duty cycle of a Medium Access Control (MAC) protocol is made up of sleep phase, wake-up phase, and listen phase. MAC protocols usually propose to optimize the duration of the wake-up and listen phases, to increase the duration of the sleep phase, thereby reducing the unwanted energy consumption of the wireless node. In this chapter, we propose an AI and machine learning (ML)-based approach, which uses a hybrid combination of Time Division Multiple Access (TDMA), Bitmap Assisted MAC (BMA), and Sensor MAC (SMAC). The machine learning layer utilizes the duty cycle in the MAC layer and generates multiple solutions for a given wireless communication. The AI layer then selects the best solution from the generated solutions by incorporating a duty cycle factor in the selection function, thereby optimizing the duty cycle of the protocol. The proposed system shows a 15% improvement in communication speed and a 10% reduction in energy consumption across multiple communications. We plan to further extend this work for rural India and apply it to real-time agricultural applications.

Data collection is a major operation in Wireless Sensor Networks (WSNs) and minimizing the delay in transmitting the collected data is critical for a lot of applications where specific actions depend on the required deadline, such as event-based mission-critical applications. Scheduling algorithms such as Time Division Multiple Access (TDMA) are extensively used for data delivery with the aim of minimizing the time duration for transporting data to the sink. To minimize the average latency and the average normalized latency in TDMA, we propose a new efficient scheduling algorithm (ETDMA-GA) based on Genetic Algorithm (GA). ETDMA-GA minimizes the latency of communication where two-dimensional encoding representations are designed to allocate slots and minimizes the total network latency using a proposed fitness function. The simulation results show that the performance of the proposed algorithm outperforms the existing state-of-the-art approaches in terms of average latency, average normalized latency, and average schedule length.

Underwater optical wireless communication (UOWC) networks are promising for many civilian and industrial applications due to underwater high-speed data transmission demand. This chapter focuses on the distributed time division multiple access (TDMA) protocol design for UOWC networks. Unlike existing research, we focus on a more general underwater communication scenario, where nodes are scattered and node mobility is considered. We propose a distributed TDMA-based medium access control (MAC) protocol, called cluster-based cross-layer multi-slot MAC (CCM-MAC), which can assign multiple slots to each node according to the slot-occupying information. When using CCM-MAC, the network is divided into clusters. Each node maintains knowledge of the cluster topology and slot-occupying information of surrounding nodes through routing techniques; by updating this knowledge, collisions can also be detected and eliminated dynamically. Analysis results are presented to evaluate the performance of CCM-MAC, and simulation results verify the benefits in comparison with existing contention-based MAC protocol.

15.3   AI-ENABLED NETWORK LAYER

The network layer is concerned with controlling of operations of the subnetwork/network. A key function of the network layer protocol is how a packet is routed from the source to the destination that may be connected over the subnetwork/network that is many hops away from the source. A standardized network routing protocol (e.g., Open Short Path First (OSPF), Border Gateway Protocol (BGP), and others) needs to be used which next hop network node is in the best situated in the path to the destination node resolving the network layer address of the destination node. In addition, routing protocol, multicast group management, network-layer information and error, and network-layer address assignment are the functions of the network layer. Note that it is the function of the payload that makes these belong to the network layer, not the function of the protocol that carries them.

Networks are complex interacting systems involving cloud operations, core and metro transport, and mobile connectivity all the way to video streaming and similar user applications. With localized and highly engineered operational tools, it is typical of these networks to take days to weeks for any changes, upgrades, or service deployments to take effect. With OSPF routing, all the packets will be delivered via the same path after quality-of-service (QOS) provisioning because it cannot perform multipath routing, and this will waste resources along the other links and lead to network congestion.

The industry has traditionally relied on hardware-centric innovations and continues to do so successfully—for instance, via photonic integration, graphene, space division multiplexing, using multi-domain optical network, and encompassing core, metro, and access networks. The various network segments typically need to work together to support multilayer services and applications, either directly or through a hybrid wireless/wireline infrastructure. Note that optical networking and its routing belongs to link layer (OSI Layer 2) while OSPF routing over the Internet Protocol (IP) network belongs to network layer (OSI Layer 3). However, the process for routing remains the same and will generally refer to OSI Layer 3 for simplicity. This diverse, dynamic, and complex mesh of networking stacks, together with future mobility constraints, necessitates smart and end-to-end service-oriented software frameworks using AI/ML/DL algorithms to augment conventional hardware advancements. [6–7]

AI/ML/DL algorithms are characterized by a unique ability to learn system behavior from past data and estimate future responses based on the learned system model. With recent improvements in computational hardware and parallel computing, features like the AI/ML/DL-enabled multipath routing along with dynamic QOS provisioning can easily be implemented.

The AI/ML/DL-based routing module constructs a sequence of tuples of the vectors representing the traffic matrix and vectors representing the routing solution. The heuristic-based module computes a heuristic-based routing solution and returns it to both data-acquisition-and-results-reporting module and the dataset warehouse.

The set of tuples in the data warehouse is identified by data analytics and by training management module to train the ML-based routing module. For example, in supervised learning, the feed-forward deep neural network (DNN) learns a mapping function that best maps input traffic matrix to multipath route output in such a way that is the best match based on a priori known labeled training dataset. Once the mapping function is learned, the AI/ML/DL-based routing module computes in real-time estimates of the routing solution, that is, multipath routes with appropriate traffic distribution to meet QOS constraints.

The proposed AI/ML/DL-based multipath routing module operates based on an algorithm, for example, feed-forward Deep DNN. The fractional values of the multipath routing output make the deep learning structure a multi-output regression model. The DNN might consist of fully connected layers out of which, say, two are hidden, one is input layer, and one is output layer. A dropout technique might be applied in both hidden layers to enhance the model robustness against overfitting. Furthermore, the adaptive moment estimation such as Adam optimization algorithm is adopted to train the deep learning network and minimize the prediction mean square error (MSE). A simplified AI/ML/DL-enabled multipath routing schematic diagram is shown in Figure 15.7.

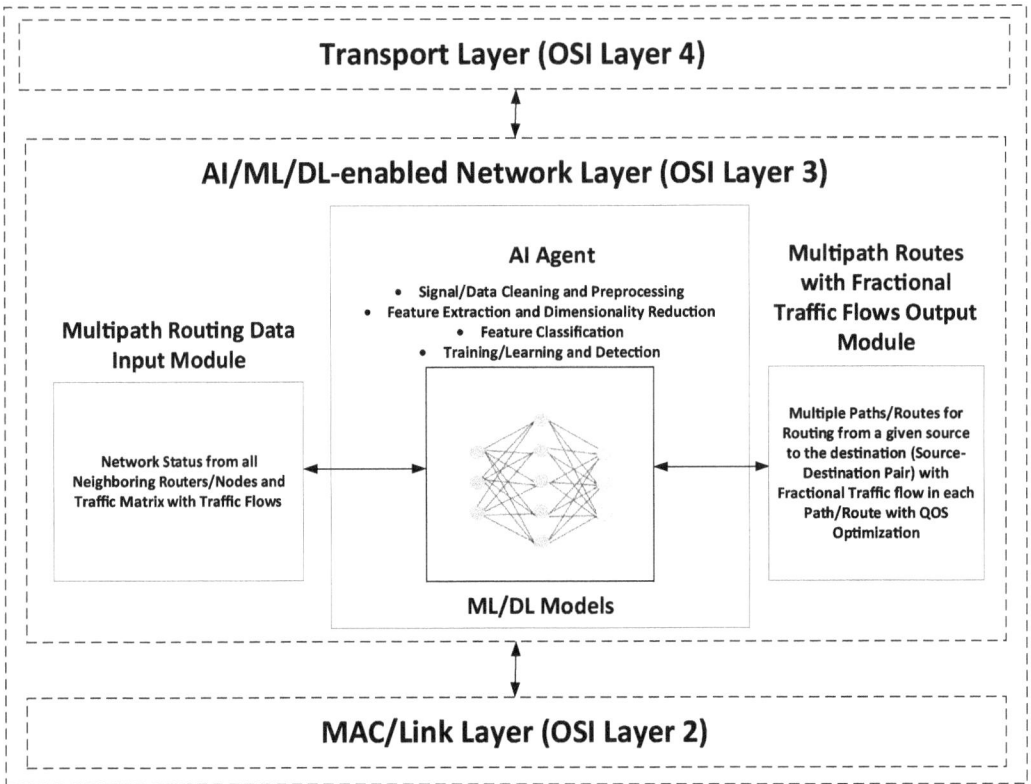

FIGURE 15.7 Simplified AI/ML/DL-Enabled Multipath Routing Schematic Diagram.

During initialization, the input module (or data acquisition and results reporting) collects the network status and traffic matrices from the router and forwards it to the heuristic-based AI/ML/DL-enabled routing module. The heuristic-based AI/ML/DL module, say, Deep Neural Network (DNN), computes a heuristic-based multipath routing solution and returns it to the multipath routing output module along with fractional traffic for each path. However, before DNN is initialized for multipath route computation, DNN needs to be trained with sufficient known labeled routing traffics by data analytics. It can be done by offline or online method.

Once the DNN is trained, the AI/ML/DL-based routing module starts generating routing solutions and reports them to the data-acquisition-and-results-reporting module, which forwards them to router. At this stage, the routing solutions received from the heuristic-based routing module are used only to update the data warehouse and only ones received from the AI/ML/DL-based routing module are forwarded to the router in real time. The DNN design can be finalized after several trial-and-error experiments. Various design versions might be considered until the best performance is achieved in both training and testing evaluations. Performance evaluations demonstrated a promising improvement in AI/ML/DL-enabled multipath OSPF in comparison to the legacy single-path open shortest path first (OSPF) protocol in terms of signaling overhead, network delay, and throughput.

In short, the entire process of AI/ML/DL-enabled routing function, say AI/ML/DL-enabled multipath-OSPF, remains the same. Choosing the specific parameters for any applications, for example, OSPF, we can tune the algorithms appropriately with the OSPF routing, we can make it AI/ML/DL-enabled multipath routing. Finally, we envisage the prospect of the intelligent physical layer and intelligent network layer. As AI continues to integrate in optical physics and electromagnetism area, the derived IRS, metamaterials, meta-surfaces, meta-lenses, and optical computing will further drive the development of IVLC. Optical computing and optical neural networks will be the focus of development. As IVLC may be deployed in 6G on a large scale, electricity-based digital signal processing will consume a lot of energy. The use of optical computing will greatly reduce consumption. Distributed channel equalization combines the existing communication system and the multilayer mechanism of neural networks, which will be an effective means of rapid deployment of IVLC. At the network layer, the main role of AI will be to reduce human intervention. It can achieve better resource scheduling and security through intelligent learning. Hopefully, it will be a thorough investigation of intelligent visible light communication and serve as a practical guide for large-scale deployment of visible light communications in future 6G networks.

15.4 AI-ENABLED TRANSPORT NETWORK LAYER

The basic function of the transport layer situated between the session (Layer 5) and the network layers (Layer 3) is to accept data from the session layer, then split into smaller packets as permitted per the network layer, and pass to the network layer ensuring that the packets will arrive correctly (for connection-oriented transport protocol such as Transmission Control Protocol (TCP) standardized in the IETF and others) at the other hand. It implies that the reliability of a given link is controlled by the transport layer through flow control, segmentation and de-segmentation, and error control. The connectionless transport protocol is used (e.g., User Datagram Protocol (UDP) standardized in the IETF and others), the reliability services of the packets are pushed to the upper application layer using some proprietary mechanisms, and flow control is also not offered by the transport layer.

TCP operation over the network is suboptimal especially for frequent transfer of large files of arbitrary sizes to network performance degradation, including queuing delays and packet drops. Solutions that predict different aspects of TCP performance, such as throughput or packet loss prediction, are largely centered on two approaches: formula-based and history-based predictions. Formula-based methods predict performance aspects using mathematical expressions that relate the predicted variable to path and end host properties such as Round-Trip Time (RTT) or the receiver's window size. In most cases, measurements for the properties are gathered using different active or passive network measurement tools.

On the other hand, history-based approaches produce a time series forecast of the desired attribute (e.g., packet loss) based on measurements derived from previous file transfers, collected either passively (e.g., through monitoring a link) or actively (e.g., by conducting file transfers of different size). Although for certain aspects of TCP performance, history-based approaches tend to be more accurate than formula-based predictions [3], existing solutions focus mostly on predicting network throughput. However, none of these methods can accurately predict end-to-end packet loss, manifested in the number of retransmitted packets in high-energy scientific application flows of arbitrary size and network throughput so that TCP can control the flow of packet transmissions and retransmissions.

AI/ML/DL-enabled TCP can automatically build a loss classifier for predicting end-to-end packet loss, manifested in the number of retransmitted packets in science flows of arbitrary size, queueing delay, end-to-end roundtrip delay, and network throughput. It will consider both formula- and history-based methods in addition to measurements of path and host attributes from previous data transfers in real time [8–9]. Machine learning algorithms like Random Forest Regression, Decision Tree Boosting, and others. A simplified AI/ML/DL-enabled TCP transport protocol schematic diagram is shown in Figure 15.8. Note that the overall AI/ML/DL process still remains the same as described in Figure 15.8.

Since retransmissions are mostly due to physical loss along the selected path and suboptimal end host network tuning, different combinations of end host properties (e.g., TCP max segment size) and path-related measurements (e.g., Round Trip Time) need to be considered in the model. There can be many numbers of collected measurements of features, which would provide an overly generalized solution with increased computational cost. However, the following set of features is selected:

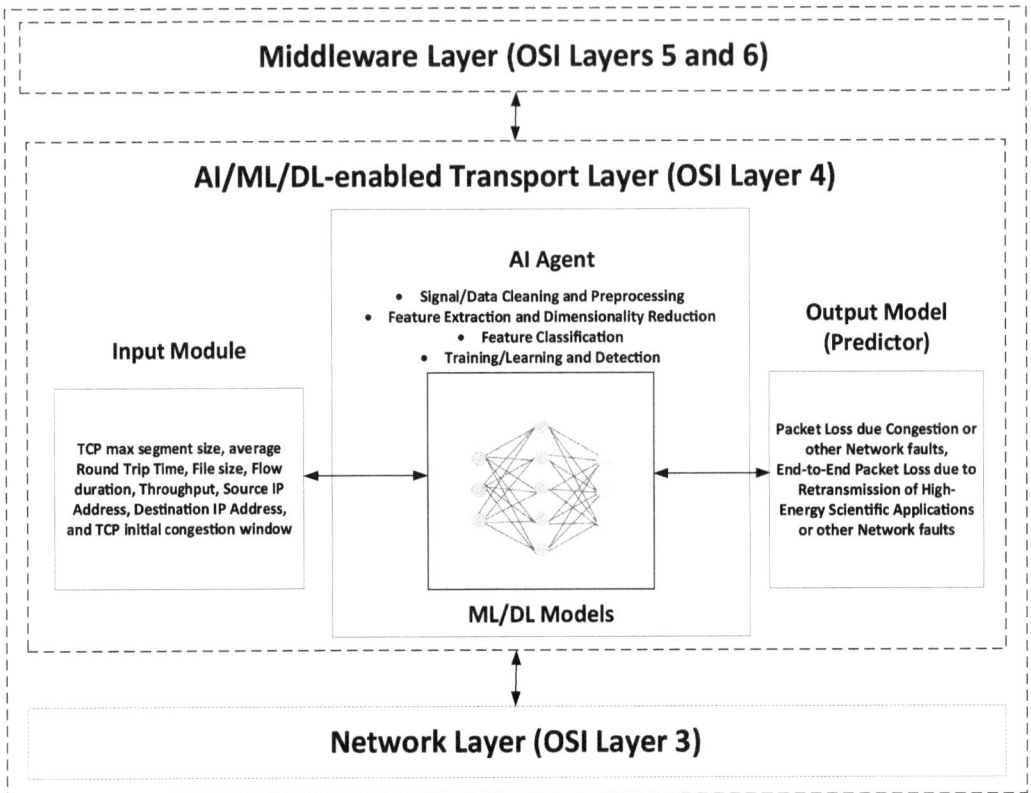

FIGURE 15.8 Simplified AI/ML/DL-Enabled TCP Transport Protocol Schematic Diagram.

TCP max segment size, average round trip time, file size, flow duration, throughput, source IP address, destination IP address, and TCP initial congestion window.

With such a classifier, the modified protocol will proceed as follows: each time a loss is detected by a triple duplicate, its cause is determined by the classification model. If the packet loss/end-to-end packet loss is classified as due to a congestion, the sender proceeds as usual (i.e., it divides its congestion window by 2). If the end-to-end packet loss is due to retransmission of high-energy scientific applications, it would allow both scientists of the high-energy scientific applications and network operators to mitigate packet loss through different host or flow reconfiguration techniques. Otherwise, it maintains its congestion window constant. The advantage of such end-to-end methods is that they do not require the support of the networks and hence they can be more easily deployed in practice.

AI/ML/DL-enabled TCP shows a significant improvement of bandwidth usage in wireless networks and no deterioration in traditional wired networks. The predictor demonstrated low training times and was able to provide accurate estimates (97– 99%) for packet retransmissions of data transfers of arbitrary sizes. The results also manifest that our solution was able to predict retransmit behavior reasonably well (66%) even for previously unseen data if training and testing datasets had similar statistics.

15.5 AI-ENABLED MIDDLEWARE LAYER

The session layer allows users on different computers/machines to establish session between them controlling the dialog. It establishes, manages, and terminates the connections between the local and remote application. It provides functions such full/half-duplex, or simplex operation, and establishes procedures for check-pointing, suspending, restarting, and terminating a session. Note that the internet protocol layer stack that does not contain the session layer, the suspending and closing function of the session of connection-oriented transport protocol (e.g., TCP) are merged with the lower transport layer (Layer 4) as TCP establishes a session (OSI Layer 5). Audio/video coding belongs to the presentation layer (OSI Layer 6). Similarly, other applications such as message format conversion will also belong to the session layer. In practice, these two layers are combined using the term middleware layer (OSI Layers 5 and 6) because the functional elements of the session layer support the applications residing in the application layer (OSI Layer 7), and all these middleware functions can be used among themselves to compose complex functionalities without obeying the hierarchy of the protocol layers as we see in OSI layers.

AI/ML/DL emerges as an important new resource-intensive workload and has been successfully applied in computer vision, speech, natural language processing, and so on. Distributed deep learning is becoming a necessity to cope with growing data and model sizes. Its computation is typically characterized by a simple tensor data abstraction to model multidimensional matrices, a data-flow graph to model computation, and iterative executions with relatively frequent synchronizations, thereby making it substantially different from map/reduce-style distributed big data computation.

Remote Procedural Call (RPC), commonly used as the communication primitive, has been adopted by popular deep learning frameworks such as TensorFlow, which uses gRPC. We show that RPC is suboptimal for distributed deep learning computation, especially on a remote direct memory access (RDMA)-capable network [10]. The tensor abstraction and data flow graph, coupled with an RDMA network, offers the opportunity to reduce the unnecessary overhead (e.g., memory copy) without sacrificing programmability and generality.

From a data access point of view, a remote machine is abstracted just as a "device" on an RDMA channel, with a simple memory interface for allocating, reading, and writing memory regions. Our graph analyzer looks at both the data flow graph and the tensors to optimize memory allocation and remote data access using this interface. The result is up to 169% improvement against an RPC implementation optimized for RDMA, leading to faster convergence in the training process.

15.6 AI-ENABLED SESSION LAYER

Remote procedure call (RPC) is middleware protocol that is a form of client–server interaction (caller is client, executor is server), typically implemented via a request–response message-passing system in a distributed environment. A computer program initiates an RPC when a program causes a procedure (or subroutine) to execute in a different address space (commonly on another computer on a shared network). RPC is coded as if it were a normal (local) procedure call, without the programmer explicitly coding the details for the remote interaction.

However, RPC is suboptimal for distributed deep learning computation. This is because (i) deep learning computation uses tensor (or multidimensional matrix) as the main data type, which consists of a plain byte array as tensor data and a simple schema as metadata specifying the shape and element type of the tensor. A tensor is often of a sufficiently large size (tens of kilobytes to megabytes) and its metadata/data sizes are often static. Using RPC for tensor data transfer does not provide evident advantage on programmability or efficiency; and (ii) using RPC typically involves memory copy to and from RPC-managed communication buffers memory region and known statically.

A simple and almost trivial interface exposes a remote machine as a "device" from a data access point of view. This "device" is connected through an AI/ML/DL-enabled RPC channel [10–11] that exposes control for parallelism. Remote memory regions can be allocated and directly accessed through this "device" interface, much like a local GPU. This maps naturally to the underlying AI/ML/DL-enabled RPC network that provides direct remote memory access.

Based on AI/ML/DL-enabled RPC, a zero-copy cross-machine tensor transfer mechanism has been designed directly on the "device" interface. This is done through a combination of static analysis and dynamic tracing on the data-flow graph of the computation to (i) figure out whether the size of each tensor that needs to be transferred across server can be statically known at the compile time, (ii) assess whether such a tensor should be allocated statically (for better efficiency) or dynamically (for reduced memory footprint), (iii) ensure allocation of the tensors on both the sending and receiving ends in the AI/ML/DL-enabled RPC memory regions, (iv) identify the source and destination addresses of tensors for AI/ML/DL-enabled RPC transfer. The experiments show that AI/ML/DL-enabled RPC techniques help TensorFlow achieve up to 169% improvement in representative deep learning benchmarks against an RPC implementation optimized for AI/ML/DL-enabled RPC. The process for AI/ML/DL-enabled RPC still remains the same as described in Figure 15.8 and OSI layer 2, 3, or 4.

15.7   AI-ENABLED PRESENTATION LAYER

Conventional audio coding technologies commonly leverage human perception of sound, or psychoacoustics, to reduce the bitrate while preserving the perceptual quality of the decoded audio signals. AI/ML/DL-enabled audio coding, for example, can decode signals more perceptually like the reference, yet with a much lower model complexity [12–13]. The proposed loss function incorporates the global masking threshold, allowing the reconstruction error that corresponds to inaudible artifacts. Experimental results show that the proposed model outperforms the baseline neural codec twice as large and consuming 23.4% more bits per second. With the proposed method, a lightweight neural codec, with only 0.9 million parameters, performs near-transparent audio coding comparable with the commercial MPEG-1 Audio Layer III codec at 112 kbps. We have not shown the detail of this because the process of AI/ML/DL-enabled audio coding still remains the same as described earlier (Figure 15.8 and OSI Layer 1).

15.8   AI-ENABLED APPLICATION LAYER

The application layer is the layer where the end-user interacts directly with the software applications. OSI standard does not define this layer leaving to the market to develop intelligent and value-added applications and services through competitive means based on market demand fueled by customers throughout the world. Of course, there are opportunities to standardize some application layer protocols in the international standard forums such as Simple Mail Transfer Protocol (SMTP) for emails, Session Initiation Protocol (SIP) for Voice/Multimedia-over-Internet Protocol (VoIP), File Transfer Protocol (FTP) for file transfers, and others. It implies that the value-added services to be built on the top of the standardized application layer protocol is so huge that some sort of interoperable products in multi-vendor environments are needed for reducing costs building scalable global communications networks. More precisely, there could be many building blocks in the application layer differencing between the application-entity and application itself. For example, application-layer functions typically include identifying communication parties, determining resource availability, and synchronizing communication. Some examples of AI-enabled applications are stated here:

- Virtual assistants like Siri and Alexa
- Recommendation systems used in e-commerce platforms
- Fraud detection in financial institutions
- Autonomous vehicles
- NLP for chatbots and customer service
- Image and facial recognition in security systems
- Medical diagnosis and health care systems

We can provide an example of how a cybersecurity application can be AI/ML/DL-enabled. In this example, we have considered a lot of things in detail what we have not shown earlier because a huge number of functions are involved. One of the examples would be to use application programming interfaces (APIs), assuming a specific function needs to retrieve from the repository. APIs are one of the best methods to do so as shown in Figure 15.9.

A user is invoking a robot using natural language processing (NLP) AI application to know details about related cyberattacks. Figure 15.9 shows all the steps that are self-explanatory and have not been repeated here. The important thing is that a lot of complexities are involved to resolve the command of users using NLP, and the API is the best way to deal with to retrieve the appropriate function to take the next step. We see that there are a couple of steps such as expression, intent, entity, conversation, action, and declaration functions to satisfy the user command. Figure 15.10 shows the detailed steps on how to find the specific API whose functions will satisfy the user command.

Knowledge Graph (KG) and Knowledge Base (KB) are two important functional entities that are used to appropriate API for resolving ambiguities of spoken words to the actual abilities/functions of the API in addition to other functional modules. Finally, ML/DL functions are invoked to find the details of the cyberattacks with all necessary parameters (e.g., when, and where cyberattacks happened, if so). Note that cyberattacks might happen at different times and at different places. It appears that there is not any specific time or place in the information received. As a result, it might include a lot of computational results by ML/DL processes stored, may be, in a database. The

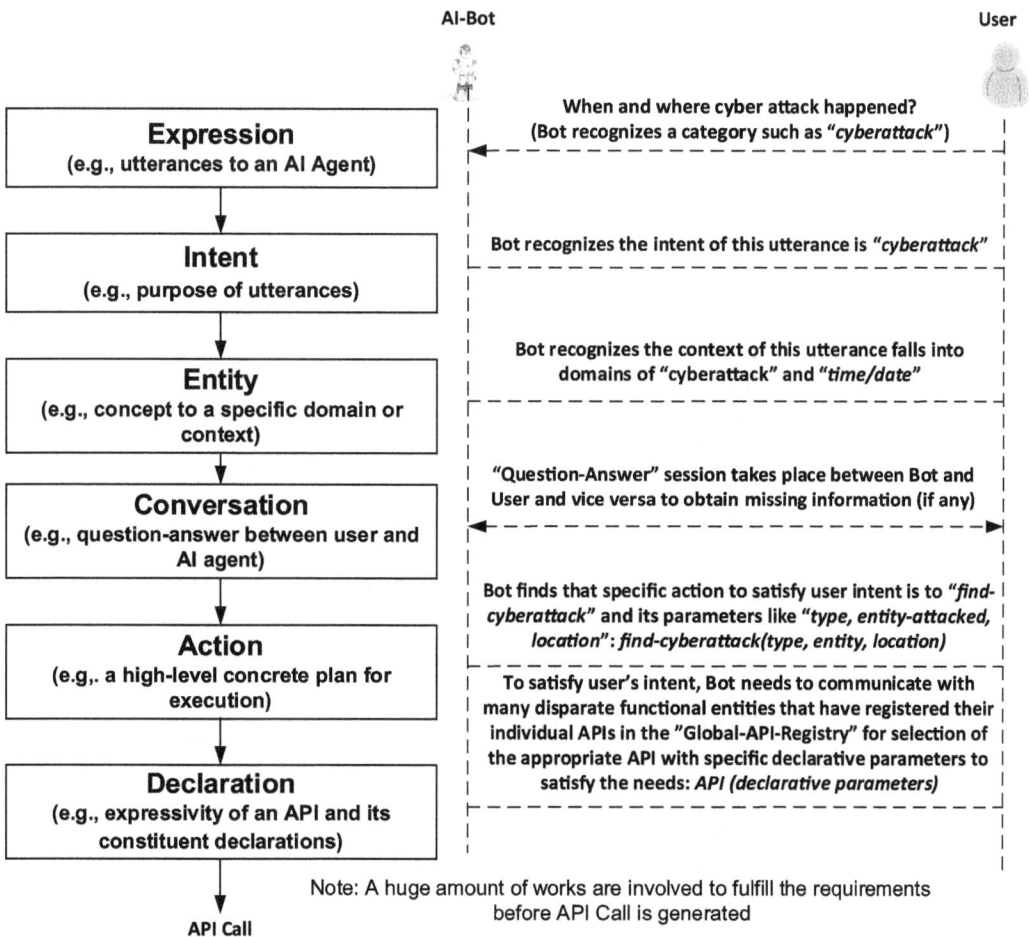

FIGURE 15.9 AI/ML/DL-Enabled Cybersecurity using Natural Language Processing (NLP) AI Application.

Choose best match declaration with maximum number of parameters fulfilled amongst candidate API declarations before the API call is generated using DTP (KG/KB) that invokes ML/DL

Feedback Loop

General Purpose NLP API

Converge Checker

KG/KB Updater

Parameter Value Groups

Entity Parameter Mapper (EPM)

Mapping Declaration-to-Parameters (DTP)

API Call
(invoking cybersecurity services)

Entities selected from the entity extractor are now matched to the parameters of the selected API using WE, PTV, and DTP of KG/KB that uses ML/DL

Candidate Declaration

API Declaration Selector

Mapping Parameters-to-Values (PTV)

Knowledge Graph (KG)/Knowledge Base (KB) Database

Expressivity (i.e. declaration) of each selected API is matched with inputted user expressions performing sematic similarity by word embedding (KG/KB) that uses ML/DL and the best match API is selected

Candidate API

API Selector

Mapping Parameters-to-Values (PTV)

General Purpose ML/DL API

Common ML/DL Infrastructure
(processes NLP requests – not shown for simplicity)

Cyberattack Detection
(result is sent back to NLP – details not shown for simplicity)

A set of viable APIs are selected to satisfy user's intent using word embedding (KG/KB) that uses ML/DL

Entities Group

Entity Extractor

Word Embedding (WE)

(also uses the same common ML/DL Services for processing of NLP services invoking API calls to ML/DL API - not shown for simplicity)

Utterances (words) are filtered to get useful words termed as "entity extraction" using Word Embedding (KG/KB) that uses ML/DL to understand basic "intent" of user

Natural Language User Expression

Note: In addition to cybersecurity, similar processes are also used for all other services

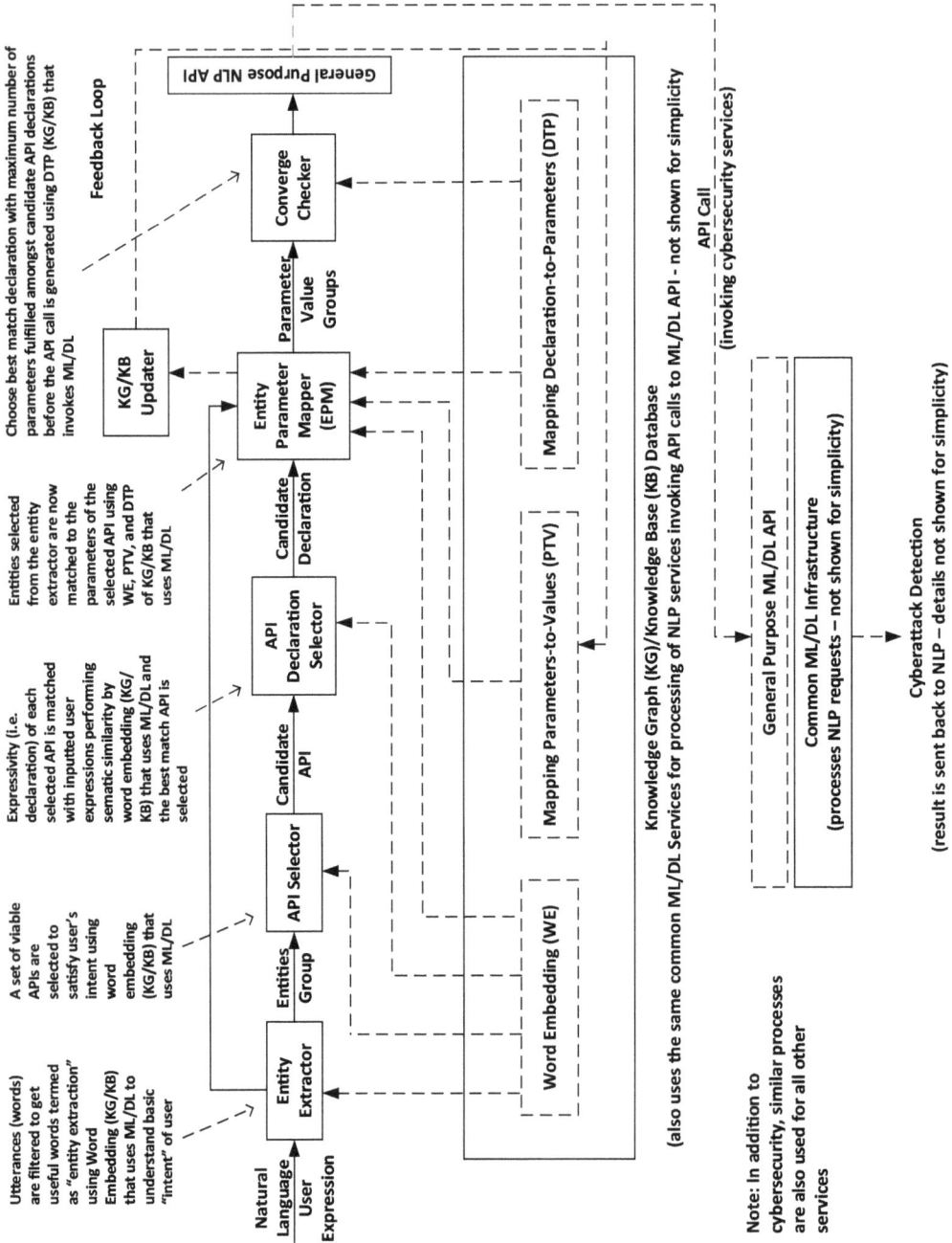

FIGURE 15.10 Cybersecurity Application enabled with NLP AI Application invoking ML/DL Algorithms.

important thing that we need to note that the basic training phase and detection phase of the AI/ML/DL-enabled cybersecurity application will remain the same as we described in Figure 15.8. However, we find an additional AI application in addition to cybersecurity such as NLP. The NLP application itself is a huge process and uses the same basic process as described in Figure 15.8. This time both NLP and cybersecurity applications that are AI-based need to be trained, and then they will be used for prediction. It indicates that a huge complexity of processing is required for making automatic AI/ML/DL-enabled cybersecurity application.

The processing and handling of high-volume data for security analysis has caused serious information overload for use in both wireless and non-wireless network. The situation becomes more problematic when the bigdata needs to be handled in mobile combat situations. Recent advances in artificial intelligence (AI) and machine learning (ML)/deep machine learning (DL) have made it possible to use these technologies for detection, prevention, response, and recovery against malware/cyberattacks [14] even in ad hoc mobile networks (MANETs) that have serious bandwidth and performance constraints. The cybersecurity analysis becomes more complicated for networks that consist of manned and unmanned ground MANETs, mobile cellular networks, unmanned aerial vehicle (UAV) networks, mobile and geostationary satellite networks, and terrestrial networks spanning across the globe. On the other hand, AI/ML/DL is becoming a primary mechanism for extracting information of big data from sensors and databases distributed across the global network. The big data models for ultra-high-dimensional problems have put tremendous pressure on AI/ML/DL methods to scale beyond a single machine due to both space and time bottlenecks. It is simply incomprehensible how security promise of instantly communicating cyber threats, patterns, and attacks can be detected in real time without the use of AI/ML/DL capabilities. At present, a variety of AI/ML/DL proprietary products have proliferated the marketplace. However, the implementation with multi-vendor products requires a standard architecture for providing interoperability and thereby offering economies-of-scale even with first-generation of machine learning products. In this respect, we are proposing a framework for AI/ML/DL-based cybersecurity network architecture.

We are describing the following: cybersecurity using AI/ML/DL, usage of AI/ML/DL in communications networks, common AI/MLDL-based applications infrastructure, cybersecurity application that uses the same common AI/ML/DL infrastructure which is also used by other applications, detailed description of the proposed common AI/ML/DL-based architecture framework including cybersecurity and some discussions about high-level APIs for ushering interoperability in multi-vendor environments until the time AI-based protocols are standardized and conclusion drawn along with future works including standardization.

15.8.1 AI/ML/DL-Enabled Cybersecurity

Cyberattacks are the key concern [5] in both types of networking environments of the 21st century: Public Internet and Military and Commercial Network. The AI/ML/DL-based cybersecurity has emerged as viable technologies especially in automating the processing of huge databases such as searching through log files for signs of compromises, detecting behavioral anomalies to find attackers on the inside or logged in with stolen credentials [14]. Although commercial enterprise users have access to multi-mega/gigabit bandwidth connections, the military tactical network has historically to operate with disconnected, intermittent, limited bandwidth (DIL) connections (e.g., MANETs) especially in mobile, dismounted combat situations.

In this chapter, we are proposing the technology-agnostic common AI/ML/DL-based architecture framework that is equally used by both cybersecurity and non-security-related applications fostering interoperability and economies-of-scale in multi-vendor environments for both military and commercial networks. Of course, features and datasets of algorithms used for security and non-security applications will definitely be different. Similarly, features and datasets of algorithms used by bandwidth- and performance-constraints communications networks will also fundamentally be different than those of the commercial networks, but all of them will be using the same architectural framework. Later (see Figure 15.2), we have shown how different algorithms that are required either for military or for commercial network can choose algorithms in accordance with their needs, but each of them will be using the same technology-agnostic common architecture.

In addition, the cybersecurity itself needs the AI/ML/DL infrastructure for both automation and faster processing of information near-real-time [6] because it is an intractable and time-intensive task to specify a set of rule-based policies, processing of log-data, and analyzing ongoing traffic flows for every conceivable scenario of each application using classical legacy security tools. A huge challenge is how to understand every possible content, their application behavior, and target environments. Another challenge is how to deploy diverse changeable security policies, patches, authentication, authorization, and other features using manual or non-automatic ways. For example, let's consider mobile vehicles of MANETs where we seek to protect each application of each device across the entire network while a MANET is only one kind of access networks, meeting all security objectives in a non-intrusive way.

AI/ML/DL technologies can also be used for generation of security policies, patches, authentication, authorization, and other features dynamically in real-time using centralized or distributive security architecture. That is, we are envisioning that the cybersecurity, like all other applications, for communications networks will also be using the technologies like machine learning, neural network, and machine vision. The difference is that they will use different features and datasets of algorithms because of meeting different objectives but is the same common technology-agnostic architecture.

15.9 SUMMARY

We have explained an important part of the AI-enabled networking protocol layers according to OSI model. We have described in detail from physical layer to the application layer. It is also termed as the networked AI. All the AI-enabled protocol layers are being made AI-enabled using all the proposed emerging AI-based networking protocols that have published over the years. It is expected that standard bodies/fora will standardize these AI-enabled protocols for providing interoperability in multi-vendor environments.

REFERENCES

[1] Wong, L. J. and McPherson, S. "Explainable Neural Network-based Modulation Classification via Concept Bottleneck Models," 2021, arXiv:2101.01239v1 [eess.SP] 4 Jan 2021.
[2] Kshirsagar, P. R. et al., "A Review on Comparative study of 4G, 5G and 6G Networks," 5th International Conference on Contemporary Computing and Informatics (IC3I), 2022.
[3] Dutta, H. and Biswas, S., "Towards Multi-Agent Reinforcement Learning for Wireless Network Protocol Synthesis," duttahr1@msu.edu, sbiswas@msu.edu.
[4] Abbasi, M. et al., "Deep Reinforcement Learning for QoS provisioning at the MAC layer: A Survey," Engineering Applications of Artificial Intelligence. 102 (2021): 104234, 2021.
[5] Pasandi, H. B. and Nadeem, T., "MAC Protocol Design Optimization Using Deep Learning," 2020, arXiv:2002.02075v1 [cs.IT] 6 Feb 2020.
[6] Liang, K. and Myers, M., "Machine Learning Applications in the Routing in Computer Networks," kul660@psu.edu, mtm387@psu.edu.
[7] Rafique, D. and Velasco, L. "Machine Learning for Network Automation: Overview, Architecture, and Applications [Invited Tutorial]," D126 J. OPT. COMMUN. NETW./VOL. 10, NO. 10/OCTOBER 2018.
[8] Giannakou, A. et al., "A Machine Learning Approach for Packet Loss Prediction in Science Flows," 2013, Elsevier.
[9] Zhang, J. et al., "WiseTrans: Adaptive Transport Protocol Selection for Mobile Web Service," Web Conference 2021 (WWW '21), 2021.
[10] Xue, J. et al., "Fast Distributed Deep Learning Over RDMA," 2019, ACM, https://doi.org/10.1145/3302424.3303975.

[11] Xue, J. et al., "RPC Considered Harmful: Fast Distributed Deep Learning on RDMA," 2018, arXiv:1805.08430v1 [cs.DC] 22 May 2018.

[12] Purwins, H. et al., "Deep Learning for Audio Signal Processing," Journal of Selected Topics of Signal Processing. 13 (2): 206–219, May 2019.

[13] Mendes, P. R. et al., "Shaping the Video Conferences of Tomorrow With AI," 2020, SBC—Sociedade Brasileira de Computação. ISSN 2596–1683.

[14] Roy, R. R. et al., "Artificial Intelligence, Machine Learning/Deep Learning-based Warfighter Cybersecurity Architecture Framework," MILCOM, 2019.

16 AI-Enabled End-to-End Network

16.1 OVERVIEW

The 21st century is witnessing a dramatic revolution with the successful application of artificial intelligence (AI) shaped by both theories and techniques for making AI-enabled products and communications networking services including cybersecurity. However, the multidisciplinary and fast-growing features make AI a field that is difficult for incorporation of capabilities that are AI-enabled and networked using a common architecture. The open and public standards like the International Organization for Standardization (ISO) and International Telecommunications Union (ITU) have assembled massive worldwide teams for developing AI standards, but it will take many years before having AI standards addressing highly technical complexity. In the meantime, products are being developed and used for services almost in all disciplines. We have analyzed many published research papers and vendors' proprietary products for the last couple of years. This innovative chapter describes how a common net-centric architecture can make every "legacy" existing capability, including cybersecurity, AI/machine learning (ML)/deep learning (DL)-enabled and networked (or network-enabled). We have considered Open Standard International (OSI) layers protocol architecture for developing AI/ML/DL common infrastructure where existing legacy protocols across the network do not change. The proposed infrastructure will provide scalability for sharing the same "common" AI/ML/DL platform resources and somewhat interoperability using the application programming interface (API) even using proprietary products in multi-vendor environments and thereby economies-of-scale.

16.2 ARTIFICIAL INTELLIGENCE AND MULTIDISCIPLINARY APPLICATIONS

The multidisciplinary applications [1] in the areas of cybersecurity, warfighters, businesses, DNA (deoxyribo-nucleic-acid) sequencing, medical, health care, high-energy physics, chemistry, biometry, movies, televisions, real-time high-quality audio-video conferencing, astronomy, railways, transportation, traffic control, robotics, inter-planet travels/rovers, auto-driving cars, hotel services, financing and banking, legal services, food-agriculture-manufacturing industry, big data, and all other disciplines have started or are poised to use artificial intelligence. In other words, all these applications are going to be AI-enabled. The ISO and ITU have been working for creation of AI standards bringing AI expertise throughout the world. A massive amount of AI items are taken considering all multidisciplinary applications. It will take decades for ISO and ITU to develop newer common standards for AI with much newer, simpler, intelligent communications protocols but with massive computation in the backend.

In the meantime, we have investigated a good amount of research publications and emerging vendors' products in the marketplace related to AI to be used with "existing" legacy networking protocols. We have analyzed the OSI protocol layers for how we can apply OSI models for AI/ML/DL. First, the multidisciplinary applications described earlier primarily belong to the OSI application layer (OSI Layer 7). Second, it is observed that all these existing legacy applications are AI-enabled using a core set of AI applications. Third, feature sets of each application could be huge, for example, a few hundreds to millions. Fourth, preprocessing, feature selection, feature classification, and evaluation need highly computation-intensive complex mathematical analysis

DOI: 10.1201/9781003499466-16

using ML/DL algorithms/models. Fifth, an AI agent of each AI-enabled application invokes one or more AI applications; in turn, AI applications invoke ML/DL algorithms/models for preprocessing, feature selection, feature classification, and evaluation. Sixth, like AI, ML/DL has a core set of algorithms/models that could be triggered in response to AI-enabled existing legacy application's specific requirements. Seventh, many open AI/ML/DL tools are being developed in the industry that are used by millions of users throughout the world. Eighth, not only existing applications, each capability of each OSI layer from Layer 1 to Layer 7 can be AI/ML/DL-enabled. Finally, OSI layers 1 through 4 are used to make each capability of the upper OSI layers from 5 through 7 networked. Our proposed common net-centric AI communications architecture is based on these technical foundations.

16.3   COMMON AI/ML/DL COMMUNICATIONS NETWORK INFRASTRUCTURE

OSI layer protocol stack (Figure 16.1) has served the communications networking extremely well for developing existing legacy open communications protocol standards and fostering interoperability in multi-vendor environments for more than four decades. The open standards have fostered the ability to build scalable products and provide services with lower cost across the whole world.

However, these legacy standardized protocols are neither AI-enabled nor AI-based. Delays in developing common standards for multidisciplinary AI including cybersecurity replacing the existing legacy protocols have caused the product vendors to develop their own proprietary products causing silos of non-interoperable costly products and services. We have analyzed that a common AI/ML/DL infrastructure or platform that acts as the middleware (OSI Layers 5 and 6) is used to make every application AI-enabled for communications networking performing AI-related computations in the backend while keeping existing "legacy" networking protocols as they are for proving interoperability across the global networks as they have been doing today. The term middleware is a fuzzy one since it has not been defined in standard bodies per se. Over decades of experience in building application software, it is seen that many applications software can be reused for building another set of value-added applications and vice versa. Industry has developed many well-known middleware.

In the same token, we have seen that a core set of AI applications and ML/DL algorithms/models can be reused to make each capability of each OSI layer, AI-enabled or AI/ML/DL-enabled and finally AI-based. That is why, we have shown in Figure 16.1 that the common AI/ML/DL infrastructure or platform belong to middleware (OSI Layers 5 and 6). The core set of common AI applications that are used as the middleware are as follows: Natural Language Processing (NLP) Platform (Content Extraction, Classification, Machine Transformation, Questions Answers, and Test Correlation), Expert System Platform, Knowledge Base (KB), Ontology, Vision Platform

OSI Layer 7	Application Layer	AI-enabled Applications
OSI Layer 6	Middleware Layer	Common AI/ML/DL Platform
OSI Layer 5		
OSI Layer 4	Transport Layer	AI-enabled Transport Layer
OSI Layer 3	Network Layer	AI-enabled Network Layer
OSI Layer 2	MAC/Link Layer	AI-enabled MAC/Link Layer
OSI Layer 1	Physical Layer	AI-enabled Physical Layer (Signal Processing)

FIGURE 16.1   OSI Layer Model and Common AI/ML/DL Platform.

(Image Processing/Machine Vision), Speech Platform (Text-To-Speech/Speech-To-Text), Robotics Platform, and other Platforms.

Similarly, the core set of ML/DL algorithms/models that also belong to middleware (Figure 16.1) are as follows [2–3]:

- **Signal/Data Preprocessing:** Noise Removal; Tokenization; Stop Words; Capitalization; Slang and Abbreviation; Spelling Correction; Stemming; Lemmatization; and others.
- **Syntactic Word Representation**: N-Gram and Syntactic N-Gram; Weighted Words: Bag-of-Words (BoW)/Beg-of-Feature (BoF); Term Frequency-Inverse Document Frequency (TF-IDF); Word Embedding-Word-to-Vector (Word2Vec): Continuous Bag-of-Words Model and Continuous Skip-Gram Model; Global Vectors for Word Representation (GloVe); Word Embedding-Contextualized Word Representations; and others.
- **Component Analysis**: Principal Component Analysis (PCA), Independent Component Analysis (ICA); Linear Discrimination Analysis (LDA); Non-Negative Matrix Factorization; Random Projection: Random Kitchen Sinks, Johnson–Lindenstauss Lemma; Autoencoder: Conventional Autoencoder Architecture and Recurrent Autoencoder Architecture; t-distributed Stochastic Neighbor Embedding (t-SNE); Belief Network; and others.
- **Logistic Regression (LR)**: Naïve Bayes Classifier; K-Nearest Neighbor (KNN): Weight Adjusted K-Nearest Neighbor Classification; Support Vector Machine (SVM) Design Tree; Random Forest; Neural Tuning Machine (NTM), and others.
- **Deep Neural Networks**: Recurrent Neural Network (RNN): Long Short-Term Memory (LSTM) and Gated Recurrent Unit (GRU); Conventional Neural Networks (CNN); Deep Belief Network (DBN); Hierarchical Attention Networks (HAN); Stochastic Gradient Descent (SGD) Optimizer; Adam Optimizer, Recurrent convolutional neural networks (RCNN); CNN with LSTM (C-LSTM); and others.

The performances of AI/ML/DL algorithms/models are evaluated using the following models that also belong to the middleware as shown in Figure 16.1 (OSI Layers 5 and 6): Confusion Matrix; Macro- and Micro-Averaging; F_β Score; Matthews Correlation Coefficient (MCC); Receiver Operating Characteristics (ROC); and Area under Receiver Operating Characteristic (ROC) Curve (AUC).

We have divided ML/DL algorithms/models into some kind of loose groups (a) through (f). In general, group (a) is good for data clearing and preprocessing. Group (b) seems to be suitable for feature extraction while group (c) is good for dimensionality reduction. However, groups (d) and (e) are used for features classification. Group (e) belongs to DL algorithms that primarily use multilayer neural networks. However, this grouping is not a restriction as algorithms of one group could be used for doing functions of other groups as well. Finally, group (f) is used for performance evaluation of AI/ML/DL algorithms/models.

The AI-enabled application, for example, cybersecurity, has an AI agent that invokes one or a set of core AI applications based on the requirements of the application (e.g., cybersecurity). For cybersecurity, the user might have to know when (time) and where (place and functional entity/resource) the cyberattack has happened in the network invoking the voice command (say, NLP AI application) to the AI bot (that is, AI agent). In turn, NLP will invoke appropriate ML/DL algorithms/models to process the user command to get the answer. The answer could mean the prediction of time (when) and where (resource) the cyberattack has happened over the communications network. The entire AI/ML/DL process can be generalized (also see Figure 7.2): Preprocessing, Feature Extraction, Dimensionality Reduction, Classification/Prediction, and Evaluation. Explainable AI (XAI) and Causal Inference with Reflection (CI-R)-based ML/DL are emerging as new AI/ML/DL technologies.

The signal/data clearing and preprocessing is used primarily to remove noise and processes that depend on the types of applications. In machine/deep learning, the initial set of actionable non-redundant features (values) that have been extracted from the data are used to meet the desired objectives such as facilitating the subsequent training/learning and generalization for using the trained models without being limited to specific cases, and in some cases, it might lead to better human understanding. The initial set of features is called the full set of features with no redundant features. Many ML/DL feature extraction algorithms are available as described earlier. However, what feature extraction algorithms need to be selected will again depend on the kind of applications. Note that there can be many feature extraction algorithms that will be suitable for a given AI-enabled application. One might use one or more feature algorithms suitable for that application. A usual case is to use as many algorithms as possible and select the best algorithm that yields superior results. So it is a huge endeavor to work with appropriate data to select the appropriate algorithms for the desired application.

Once the full set of non-redundant features is available, we must identify the useful features and then select the final feature set from those useful feature-sets for the application. It involves trial and error to reduce the dimensionality of the features that are the useful minimum feature set. Again, many general dimensionality reduction ML/DL algorithms are available as explained earlier. Dimensionality reduction is mostly used for improving computational time and reducing memory complexity.

The most important step of the text classification pipeline is choosing the best feature classifier. Without a complete conceptual understanding of each algorithm, one cannot effectively determine the most efficient model for a feature classification application. The most popular techniques for feature classification are listed earlier in groups (d) and (e).

Non-parametric techniques have been studied and used as classification tasks such as k-nearest neighbor (KNN). Support Vector Machine (SVM) is another popular technique which employs a discriminative classifier for document categorization. This technique can also be used in all domains of data mining such as bioinformatics, image, video, human activity classification, safety, and security. This model is also used as a baseline for many researchers to compare against their own work to highlight novelty and contributions [2].

Tree-based classifiers such as decision tree and random forest have also been studied with respect to document categorization. Each tree-based algorithm will be covered in a separate subsection. In recent years, graphical classifications have been considered as a classification task such as conditional random fields (CRFs). However, these techniques are mostly used for document summarization and automatic keyword extraction.

Lately, deep learning approaches have achieved success in surpassing results in comparison to previous machine learning algorithms on tasks such as image classification, natural language processing, face recognition, etc. The success of these deep learning algorithms relies on their capacity to model complex and nonlinear relationships within data. The key is that one must select a minimum set of algorithms to test their applications for better classification before making the final decision.

The final part of the classification pipeline is evaluation for any applications. Understanding how a model performs is essential to the use and development of, say, cybersecurity, classification methods. There are many methods available for evaluating supervised techniques. Accuracy calculation is the simplest method of evaluation but does not work for unbalanced datasets. Unlike classical methods, the ML/DL technique requires the results should be tested by a set of two or more performance evaluation algorithms through comparisons before selecting the final best prediction.

The ML/DL algorithms/models could be used by many different applications. However, values of the parameters (that is, features) and their usages to interpret the results will be completely different depending on specific applications. The predicted results will be very specific to the respective applications. We have not discussed the evaluation of results comparing different applications in detail for the sake of brevity.

Note that we have not specifically described the training of AI/ML/DL algorithms/models (also see Figure 7.2). In training, known and labeled data of known applications is used so that the models can predict the actual result for that application (e.g., loss parameter exceeded the certain threshold for different kinds of cybersecurity attacks) while in actual real-world application unknown and unlabeled data is used as the input for prediction. The training data is very crucial because prediction is as good as trained data.

16.4 AI/ML/DL NETWORKING ARCHITECTURE

We have found that the common AI/ML/DL infrastructure that resides in the backend can provide services to all OSI layers to make every capability of each OSI layer AI-enabled including cybersecurity. As a result, we can use the same common AI/ML/DL platform to offer for AI-enabling services. Figure 16.2 depicts a high-level conceptual view of the common AI/ML/DL infrastructure logical networking architecture for both client and server where the AI/ML/DL platform server is offering logically the same common services to make each capability of each OSI layer AI-enabled for both client and server separately.

The important things are that vendors are making different capabilities AI-enabled in different layers separately and creating individual silos without reusing any common AI capabilities (e.g., same algorithm but different hyper-parameters in accordance with different applications capabilities needs). As a result, existing AI-solutions lack scalability. Based on logical networking architecture, a high-level network implementation architecture is shown in Figure 16.3. All the legacy servers of the server complex are AI-enabled using a common AI/ML/DL platform server without replacing the existing legacy software or hardware.

The key is that all primitive communications protocols and signal processing can be enhanced and optimized creating new lightweight intelligent protocols and signaling processing schemes allowing the high-intensive complex computation at the backend common AI/ML/DL infrastructure server for both client and server (something like cloud/fog computing). If those protocols are newly designed taking advantage of the AI/ML/DL capabilities, there needs to be new standardization by the standard bodies replacing the existing legacy standard protocols as depicted in Figure 16.4.

As we have explained earlier, it will take perhaps a decade or more to have those new protocol standards from the standard bodies. Of course, we are offering these new insights to the standard bodies how a common AI/ML/DL platform can be used in creating a new set of protocols to simplify the standards creation. In the interim, however, we are offering a solution for how we can provide a common AI/ML/DL infrastructure as a service for both client and server (Figures 16.2 and 16.3) to make each capability of each OSI layer AI/ML/DL-enabled including cybersecurity.

It implies that all existing protocols will remain the same to the outside world including cybersecurity, but AI/ML/DL services will be used as backend services to process each capability of each protocol of each OSI layer to make each of those capabilities AI-enabled. In this way, the present interoperability using legacy protocols over the Internet and/or warfighter networks remain intact.

It is clearly seen that this configuration provides a huge economies-of-scale since none of the existing legacy application servers need to be equipped with AI/ML/ML platform-specific hardware and/or software duplicating the same costly things in each server or client unnecessarily. If the application servers are procured from different vendors, the common AI/ML/DL platform application programming interface (API) can be used for offering interoperability in a multi-vendor environment. In each client side (that is, AI-enabled device), the same common AI/ML/DL infrastructure can be used for all OSI layers as depicted in Figures 16.2 and 16.3 designing the client devices such as smartphones, intelligent laptops, and other AI-enabled devices.

16.5 END-TO-END AI-ENABLED OSI LAYER USING
COMMON AI/ML/DL INFRASTRUCTURE

The description of this section is somewhat like what has been explained in Section 15. We briefly recap here in the context of end-to-end networking. All existing classical applications of OSI layers 1 through 7 (Figures 16.1, 16.2, and 16.3) can be made AI/ML/DL-enabled using the same common AI/ML/DL infrastructure described in Sections 16.3 and 16.4. The overall AI/ML/DL processing

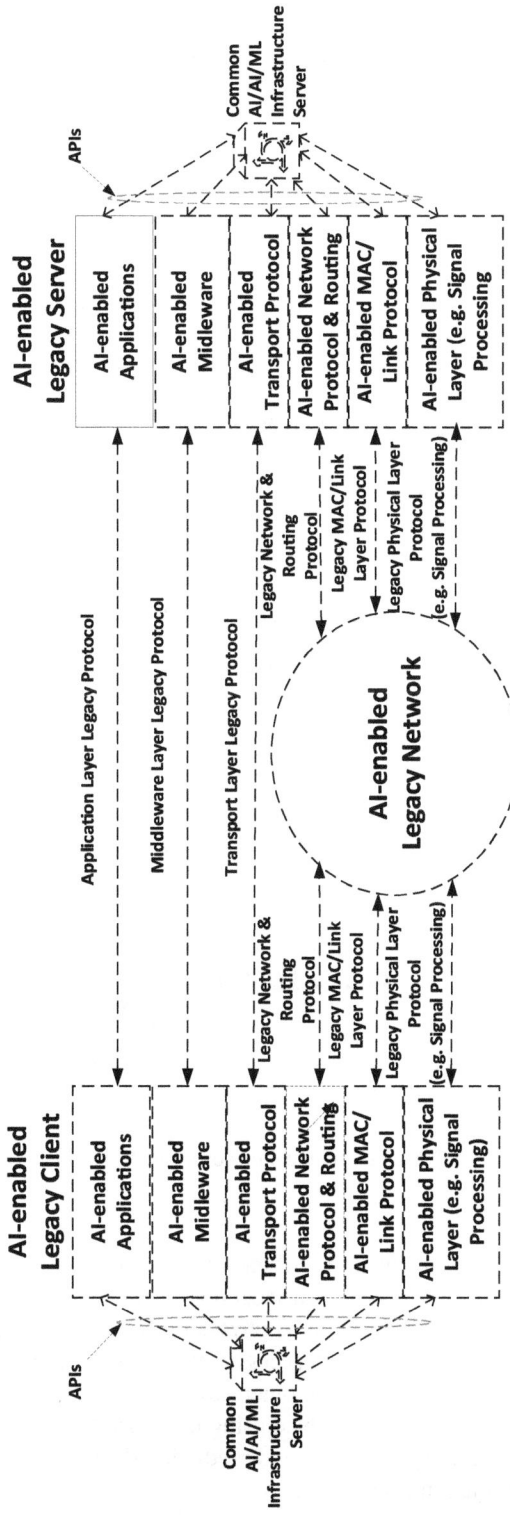

FIGURE 16.2 High-Level View of Common AI/ML/DL Infrastructure Logical Networking Architecture making Legacy Client and Server making AI-enabled.

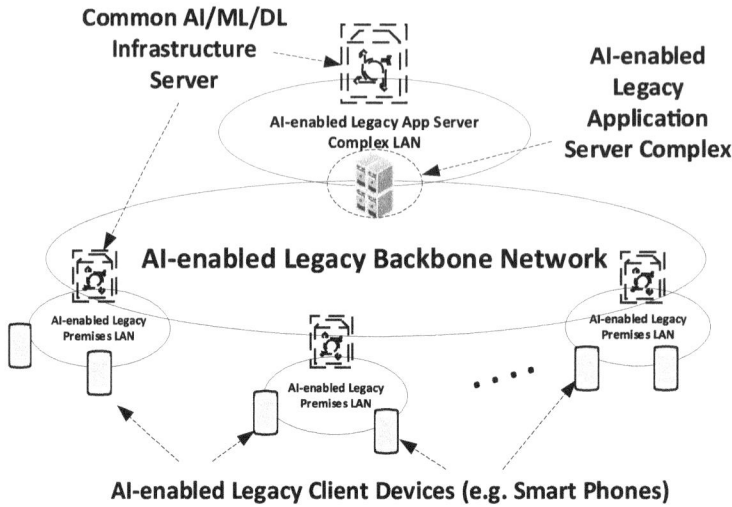

FIGURE 16.3 AI-enabled Legacy Clients, Servers, Premises, LANs, and Backbone Network.

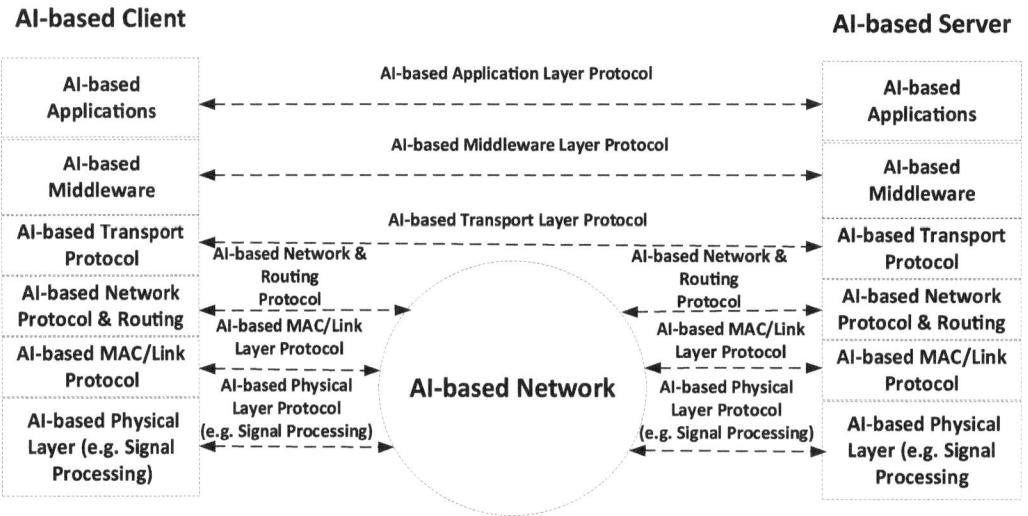

FIGURE 16.4 AI-based Server and Client communications over the AI-based Network.

technique remains the same for each capability of each OSI Layer as depicted in Figure 7.2. The physical layer is responsible for transmission and reception of raw bits over a communication link or channel. The physical layer link/channel can be physically wired or wireless. The traffic multiplexing, modulation scheme, data rate, channel access method, physical connectors, simplex, full/half-duplex, timing, frequency, voltage level, and others are the functions of the physical layer [4].

Interestingly, if a modem is AI/ML/DL-enabled using the processes (signal acquisition, preprocessing, feature extraction, feature classification, and prediction/decision) as depicted in Figure 7.2, its basic functions will still remain the same, but its performance such as bit-error-rate (BER) will be almost negligible to normal, and the channel capacity will be almost nearer to normal Shannon Channel Capacity.

The convolutional neural network (CNN) and residual neural network achieved accuracies of over 80–90% especially with the knowledge of physics of the wireless channel. Note that the processes of AI/ML/DL remain the same, but training with specific signal/data using specific information chosen from a common set of software repositories of algorithms/models requires tuning their parameters to that specific signal/data for physical layer (OSI Layer 1). Finally, the simplified representation of AI/ML/DL-based modem over the communication channel is shown in Figure 15.3 as an example.

We can now argue that a physical modem device might be designed with specific AI/ML/DL embedded hardware as a standalone hardware. The other alternative would be to have the common AI/ML/DL computing processes for all OSI Layers (layers 1 to 7) in devices integrated with modem like smartphones, warfighter radio waveform radios, and application servers (see Figures 16.2 and 16.3).

The **medium access control** (MAC) protocol in wireline/wireless networks comes with multiple levels of heterogeneity. The current best practice for programming MAC logic in an embedded wireless node is to implement known protocols such as ALOHA, Slotted-ALOHA, carrier sense multiple access (CSMA), and CSMA with collision avoidance (CSMA-CA) (i.e., Wi-Fi, etc.) depending on the available lower layer hardware support. An example is when nodes without carrier sensing abilities run the AOLHA family of protocols [5], their performance starts degrading due to collisions when the application traffic load in the network exceeds an optimal level. Such problems are further compounded in the presence of various forms of heterogeneity in terms of Acknowledgement (ACK), No-ACK, Fragmentation/Aggregation, Fragmentation Threshold, Back-off, Request-To-Send (RTS)/Clear-To-Send (CTS), Network Allocation Vector (NAV), NAV Timeout, Contention Window (CW), Traffic Load, Topology, and Node-specific Access priorities.

The important reason for such a performance gap is that the nodes are statistically programmed with a protocol logic that is not cognizant of time-varying load situations and such heterogeneities. The AI/ML/DL-enabled MAC allows wireless nodes to learn how to detect such conditions and change transmission strategies in real time for maximizing the network performance, even under node-specific access prioritization. Two ways of AI-enabled MAC protocol can be enhanced: modest performance improvement without modifying the existing legacy MAC protocol and optimized the performance creating new MAC protocol. The latter will take decades to be standardized while the former can be done now, and an API will provide interoperability in a multi-vendor product environment.

The wireline/wireless nodes with an AI/ML/DL-enabled MAC can use more simplified features to run a new rudimentary MAC logic [5] without relying on carrier sensing for example and other complex features from the underlying hardware in the future when standardized. In other words, the developed mechanisms could even be more suitable for very simple transceivers that are found in low-cost wireless sensors, Internet-of-Things (IoT), and other embedded devices. Because machine learning algorithms allow network nodes to individually adjust transmit probabilities so that self-collisions and inter-collisions are reduced without embedding any of those features in the MAC protocol itself. As a result, the nodes can attain the theoretical maximum throughput and sustain it for higher loading conditions when AI-based protocols are standardized (Figure 16.4). However, the detailed discussion of this is out of scope of this book.

The **network layer** is concerned with controlling operations of the subnetwork/network, for example, in the Internet Protocol (IP). A key function of the network layer protocol is how a packet is routed from the source to the destination that may be connected over the subnetwork/network many hops away from the source. A standardized network routing protocol (e.g., Open Short Path First (OSPF), Border Gateway Protocol (BGP), and others) needs to be used to determine which next hop network node best situated in the path to the destination node resolving the network layer address of the destination node. AI-enabled OSPF can process the next-hop route calculation in a much efficient way knowing both history and the present network condition keeping existing OSPF protocol unchanged.

With OSPF routing, all the packets will be delivered via the same path after quality-of-service (QOS) provisioning because present OSPF routing protocol cannot perform multipath routing, and this will waste resources along the other links and lead to network congestion. AI-enabled OSPF can be further optimized by the routing using multipath routing dynamically [6], but it will require changing the existing OSPF protocol by the standard body like International Engineering Task Force (IETF). The discussion for enhancements of the OSPF protocol itself making it AI-enabled is out of scope of this paper.

The basic function of the **transport layer** situated between the session (Layer 5) and the network layer (Layer 3) is to accept data from the session layer, then split into smaller packets as permitted per the network layer, and pass to the network layer ensuring that the packets will arrive correctly, for the connection-oriented transport protocol such as Transmission Control Protocol (TCP) standardized in the IETF, at the other hand. It implies that the reliability of a given link is controlled by the transport layer through flow control, segmentation and de-segmentation, and error control.

TCP operation over the network is suboptimal [7] especially for frequent transfer of large files of arbitrary sizes to network performance degradation, including queuing delays and packet drops. Solutions that predict different aspects of TCP performance such as throughput or packet loss prediction are largely centered on two approaches: formula-based and history-based predictions.

AI/ML/DL-enabled TCP [7] can automatically build a loss classifier for predicting end-to-end packet loss, manifested in the number of retransmitted packets in bandwidth-intensive flows (e.g., high-energy physics applications) of arbitrary size, queueing delay, end-to-end roundtrip delay, and network throughput. It will consider both formula- and history-based methods in addition to measurements of path and host attributes from previous data transfers in real time using machine learning algorithms like Random Forest Regression, Decision Tree Boosting, and others. AI/ML/ DL-enabled TCP shows a significant improvement of bandwidth usage in networks (especially for wireless) and no deterioration in traditional wired networks.

Remote procedure call (RPC) is a **middleware protocol** that is a form of client–server interaction (caller is client, executor is server), typically implemented via a request–response message-passing system in a distributed environment. However, RPC is suboptimal [8] for distributed deep learning computation. This is because (i) deep learning computation uses tensor (or multidimensional matrix) as the main data type, which consists of a plain byte array as tensor data and a simple schema as metadata specifying the shape and element type of the tensor.

A simple and almost trivial interface exposes a remote machine as a "device" from a data access point of view. This "device" is connected through an AI/ML/DL-enabled RPC [8] channel that exposes control for parallelism. Remote memory regions can be allocated and directly accessed through this "device" interface, much like a local graphical processing unit (GPU). This maps naturally to the underlying AI/ML/DL-enabled RPC network that provides direct remote memory access. However, it requires development of a new standardized AI-enabled RPC, and discussion of this is omitted here because we are not discussing the development of a new AI-enabled RPC.

The **application layer** is the layer where the end-user interacts directly with the software applications in the form client-server communications architecture. OSI standard does not define this layer leaving the market to develop intelligent and value-added applications and services, including cybersecurity through competitive means based on market demand fueled by customers throughout the world. Of course, there are opportunities to standardize some application layer protocols in the international standard forums such as SMTP for emails, Session Initiation Protocol (SIP) for Voice-over-Internet Protocol (VoIP), File Transfer Protocol (FTP) for file transfers, and others. It implies that the value-added services to be built on the top of the standardized application layer protocol is so huge that some sort of interoperable products in multi-vendor environments are needed for reducing costs for building scalable global communications networks. We can make SMTP, SIP/VoIP, FTP, and others AI/ML/DL-enabled as we discussed earlier in the case of AI-enabled cybersecurity. We have not elaborated this discussion further for the sake of brevity.

16.6 NETWORKED AI-ENABLED APPLICATIONS

OSI Layers 1 through 4 are responsible for making an AI-enabled application networked. Note that OSI layers 1 through 4 do not need to be AI-enabled to make an AI-application networked (or network-enabled). The lower OSI layers need to be invoked to make the AI/ML/DL-enabled cyber-security application networked or network-enabled (Figures 16.3 and 16.4). If the lower OSI layers are AI/ML/DL-enabled, the performances of those become almost near normal. For instance, if the physical layer channel becomes AI/ML/DL-enabled, the channel capacity might approach the maximum ideal Shannon throughput capacity bound. Similar is the case for all OSI Layers 2 (MAC/Link), 3 (Network/Routing), and 4 (Transport). The overall end-to-end network throughput capacity will increase dramatically nearer to normal if those OSI Layers 1 through 4 are AI/ML/DL-enabled.

16.7 SUMMARY

We have described how a common AI/ML/DL infrastructure can make each capability including cybersecurity of all OSI layers AI/ML/DL-enabled and networked. We have clearly articulated how the proposed infrastructure will provide scalability for sharing the same "common" AI/ML/DL platform resources and interoperability using the API even using proprietary products in multi-vendor environments and thereby economies-of-scale. We have only described how each capability of each OSI layer can be AI-enabled and networked for improving performances without changing the existing standardized protocols. How performances of each protocol of each OSI layer could be fully optimized using more simplified new intelligent protocol that requires standardization will be addressed in the future.

REFERENCES

[1] Liu, J. et al., "Artificial Intelligence in the 21st Century," IEEE Access (accepted pre-publication), February 2018. https://www.researchgate.net/publication/324023933_Artificial_Intelligence_in_the_21st_Century.
[2] Kowsari, K. et al., "Text Classification Algorithms: A Survey," Information, MDPI, https://arXiv:1904.08067v5 [cs.LG] 20 May 2020.
[3] Wang, W. et al., "A Survey on Applications of Model-Free Strategy Learning in Cognitive Wireless Networks," https://arXiv:1504.03976v2 [cs.NI] 9 Feb 2016.
[4] Zhou, R. et al., "Deep Learning for Modulation Recognition: A Survey with a Demonstration," IEEE Access, February 2020, https://ieeexplore.ieee.org/stamp/stamp.jsp?tp=&arnumber=9058656.
[5] Dutta, H. and Biswas, S., "Towards Multi-Agent Reinforcement Learning for Wireless Network Protocol Synthesis," https://arxiv.org/ftp/arxiv/papers/2102/2102.01611.pdf.
[6] Liang, K. and Mitchel Myers, M., "Machine Learning Applications in the Routing in Computer Networks," Pennsylvania State University Pennsylvania State University, https://arxiv.org/ftp/arxiv/papers/2104/2104.01946.pdf.
[7] Xia, J. et al., "Rethinking Transport Layer Design for Distributed Machine Learning," ACM, APNet '19, 17–18 August 2019, https://doi.org/10.1145/3343180.3343186.
[8] Xue, J. et al., "Fast Distributed Deep Learning over RDMA," ACM, EuroSys '19, 25–28 March 2019, https://doi.org/10.1145/3302424.3303975.

17 AI-Enabled Peer-to-Peer Network

We are now describing AI-enabled Peer-to-Peer (P2P) network protocol layer architecture. Like AI-enabled network, AI-enabled P2P network will also all of its networking protocols from Layer 1 to 7 will be AI-enabled. Figure 17.1 depicts the AI-enabled Peer-to-Peer Network protocol architecture layer.

We will discuss the AI-enabled P2P applications such as AI-enabled P2P Artificial Intelligence applications that consist of Natural Language Processing (NLP) Platform (Content Extraction/Classification/Machine Transformation/Questions Answering/Test Correlation), Expert System Platform, Knowledge Base (KB), Ontology, Vision Platform (Image Processing/Machine Vision),

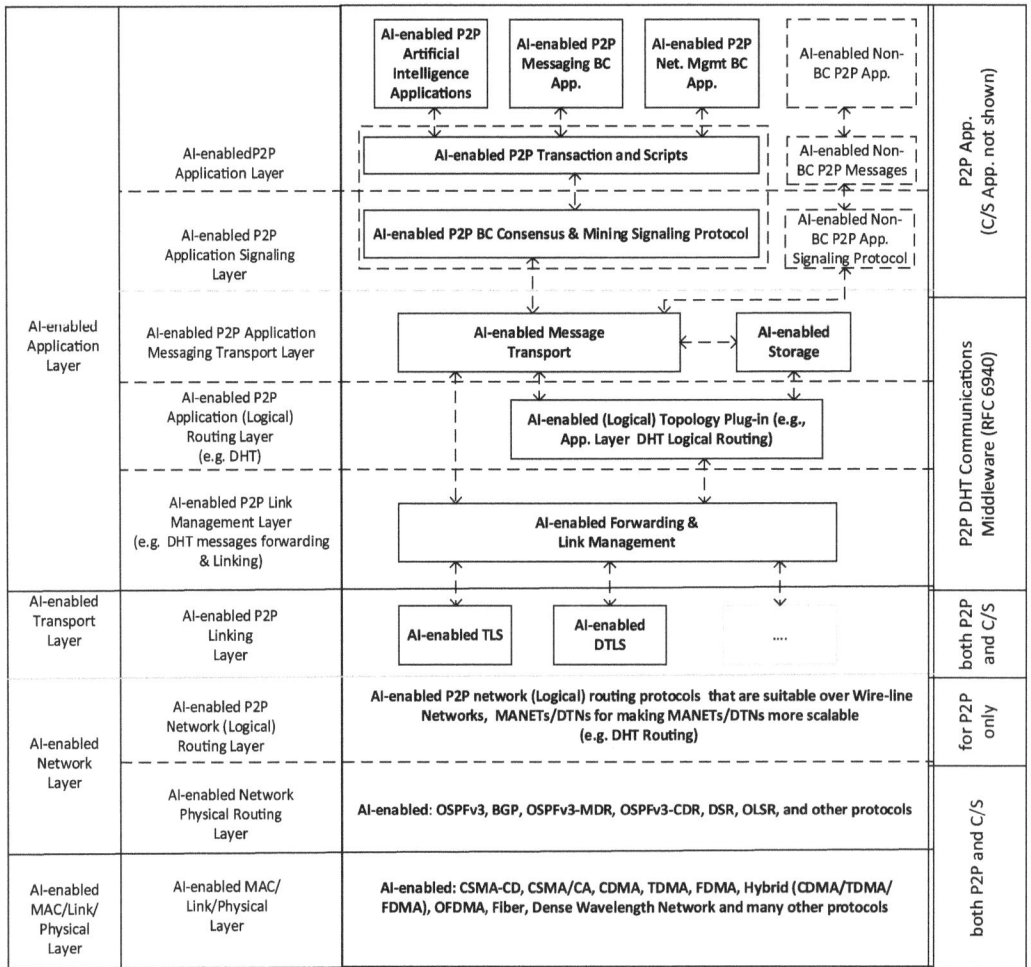

BC = Blockchain

FIGURE 17.1 AI-enabled P2P Network Protocol Architecture.

DOI: 10.1201/9781003499466-17

Speech Platform (Text-To-Speech/Speech-To-Text), Robotics Platform, and other Platforms. Each of these applications has huge functionalities and features, and AI can analyze and predict the outcome (objectives) automatically. Note that AI-enabled applications can invoke ML/DL algorithms (not shown in this figure) to meet the requirements of the applications. Similarly, AI-enabled P2P blockchain (BC) can also use AI for risk management, automatic decision-making, predictive analysis, fraud detection, portfolio management, real-time monitoring, generating data-driven insights, and many other things if applications demand to do so.

In the same token, lower echelon of sub-applications such as AI-enabled P2P Transaction, and AI-enabled consensus and mining signaling protocol, AI-enabled message transport, AI-enabled storage, AI-enabled (logical) topology plug-in, that is, application layer distributed hash table (DHT)-based application layer (logical) routing, AI-enabled forwarding & link management [1] of the application layer will also invoke AI/ML/DL capabilities as needed.

Earlier in Chapters 14 and 15, we have AI-enabled middleware, transport protocols, routing protocols, link/MAC protocols, and physical layer signal processing. The same is also applicable for AI-enabled P2P networks. In addition, there can be AI-enabled, AI-enabled Pure P2P, AI-enabled Hybrid P2P, and AI-enabled Centralized P2P networks.

17.1   SUMMARY

In this chapter, we discussed AI-enabled P2P network protocol layer architecture. Note that the client/server (C/S) network is a subset of P2P network architecture.

REFERENCE

[1] RFC 6940: REsource LOcation and Discovery (RELOAD) Base Protocol, Standard Track, January 2014.

18 Artificial Intelligence-Enabled 5G Network

18.1 OVERVIEW

Fifth-generation (5G) cellular networks are state-of-the-art wireless technologies revolutionizing all wireless systems. The fundamental goals of 5G are to increase network capacity, improve data rates, and reduce end-to-end latency. Therefore, 5G can support many devices connected to the Internet and realize the Internet-of-Things (IoT) vision. Though 5G provides significant features for mobile wireless networks, some challenges still need to be addressed. Although 5G provides valuable capabilities for mobile wireless networks, specific issues still need to be resolved. We are introducing 5G technology thoroughly, detailing its needs, infrastructure, features, and difficulties. The fifth-generation (5G) wireless network has enabled about 20 Gigabytes per second peak data rate, which is about 20 times more capacity than that of 4G technology (see Table 18.1). AI-based applications are being used in 5G network although there is no AI-enabled router or switching technology is being used.

It can be seen from Table 18.1 that the 5G is poised to provide massive connectivity, reliability and low latency, and high mobile bandwidth. Another key capability of 5G is to use software-based services hosted on reusable network infrastructure platforms. Network service functions (NSFs) based on market segments, virtualization services, and other services need to be provided through automation which implies that AI/ML/DL technologies need to be used for implementing these services. Currently as it stands, AI is designed as an external network function (NF) which is an after-the-fact service and not as the core network function. 5G improves on the 4G services over several axes:

- **Enhanced Mobile Broadband (eMBB):** Higher data rates are specified. For the downlink, up to 50 Mbps are offered for outdoor and 1 Gbps for indoor (5GLAN), with half of these values available for the uplink. Several case studies have been under consideration, among them is aviation—where eMBB is helping deliver a bit rate of 1.2 Gbps to an airborne flight.
- **Critical Communications (CC) and Ultra Reliable and Low Latency Communications (URLLC):** In some contexts, extremely high reliability is expected. For instance, for remote control of process automation, a reliability of 99.9999% is expected, with a user-experienced data rate up to 100 Mbps and an end-to-end latency much less than 50 milliseconds. This is provided through Edge Computing capability.
- **Massive Internet-of-Things (mIoT):** Several scenarios require the 5G system to support very high traffic densities of devices. The Massive Internet-of-Things requirements include the operational aspects that apply to the wide range of IoT devices and services anticipated in the 5G timeframe.
- **Flexible Network Operations:** These are a set of specificities offered by the 5G system, as detailed in the following sections. It covers aspects such as network slicing, network capability exposure, scalability, and diverse mobility, security, efficient content delivery, and migration and interworking.

This diversity of requirements, associated to the different categories of usage described earlier, enables the 5G system (5GS) to be useful to a new set of markets aka "Verticals," including

 DOI: 10.1201/9781003499466-18

TABLE 18.1

4G and 5G Technology Key Performance Parameters

Key Performance Parameters (KPPs)	4th Generation (4G) Wireless Technology	5th Generation (5G) Wireless Technology	Remarks
Latency	10 milliseconds	Less than 1 millisecond	Some AI-based applications are being used in 5G network, but no AI-based routers or switches in the network.
Data traffic	7.2 exa-bytes per month (2021)	50 exa-bytes per second (2021)	
Peak data rates	1 gigabit per second	20 gigabytes per second	
Available spectrum	3 GHz	30 GHz	
Connection density	100,000 thousand connections per kilometer square	1 million connections per kilometer square	

automotive, rail and maritime communications; transport and logistics; discrete automation; electricity distribution; public safety; health and wellness; smart cities; media and entertainment. In addition to the new 5G-specific services, the 5G system supports almost all the 4G LTE ones [1–9] and mobility between a 5G core network and a 4G core network (EPC) is supported, with minimum impact on the user experience.

Schematically, the 5G system uses the same elements as the previous generations: a User Equipment (UE), itself composed of a mobile station and a USIM, the radio access network (NG-RAN) and the core network (5GC), described later. The main entity of the NG-RAN is the gNB, where "g" stands for "5G" and "NB" for "Node B," which is the name inherited from 3G onwards to refer to the radio transmitter. The radio interface is named "NR-Uu" for similar reasons, although with divergences: here, "5G" is indicated by "NR" (for "New Radio") and Uu is also a name inherited from previous generations. The gNB may be further split into a gNB-Central Unit (gNB-CU) and one or more gNB- Distributed Unit(s) (gNB-DU), linked by the F1 interface (Figure 18.1).

To support balancing of computational resources between CUs and DUs for gNBs, **e.g., in a telecommunications environment comprising centralized and edge clouds, the F1 interface as well as the Xn-C interface need to carry information about computational resource usage, such as for example CPU, memory, and network interface utilization. Figure 18.1 illustrates the 5G-RAN architecture and the CU-DU split of functionalities.

The 5G core architecture relies on a "Service-Based Architecture" (SBA) framework, where the architecture elements are defined in terms of "Network Functions" (NFs) rather than by "traditional" Network Entities. Via interfaces of a common framework, any given NF offers its services to all the other authorized NFs and/or to any "consumers" that are permitted to make use of these provided services. Such an SBA approach offers modularity and reusability.

18.2   5G RADIO ACCESS NETWORK

We are presenting a high-level view of some possible 5G radio access network (RAN) configurations (Figure 18.2), and the core network as 6G networks are still in the research stage. The 5G core network employs 5G applications including some AI-based applications external to the core network function (CN), but RAN or core network routers or switches are not AI-based. Optimal performance of the CN, which is split into the control plane (CP) and user plane (UP), cannot be overstated. The CP consists of numerous vital NFs which must conduct significant control signaling and specialized tasks in an extremely short period of time to ensure that the end-to-end (E2E) quality-of-service (QoS) requirements are met. Otherwise, QoS flows, and user sessions cannot be established in time, thus hampering the UP routing from the RAN to the CN and then external networks. Consequently, it can even be argued that the CP latency constraints are perhaps among the

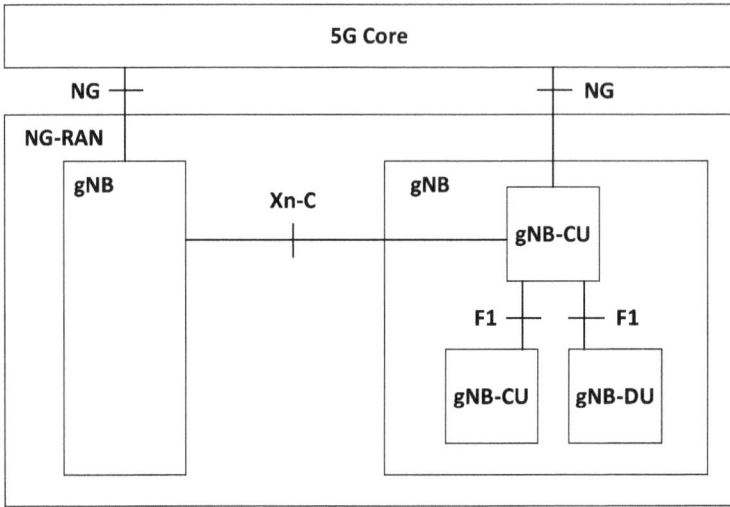

FIGURE 18.1   5G RAN Simplified Architecture.

most important and stringent constraints. 5G communications network capabilities are In-Coverage Mode, 5G Relay-Mode, and 5G Standalone.

Note that 3GPP defines RAN in the following groups:

- **RAN1**—Radio Layer 1: Physical layer.
- **RAN2**—Radio layer 2 and Radio layer 3 Radio Resource Control.
- **RAN3**—Universal Terrestrial Radio Access Network (UTRAN)/Evolved UTRAN (E-UTRAN)/Next Generation RAN (NG-RAN) including Cloud RAN (C-RAN) architecture and related network interfaces.
- **RAN4**—Radio Performance and Protocol Aspects.
- **RAN5**—Mobile terminal conformance testing.

The study has just begun, and at the time of writing we can only provide initial considerations. According to the mandate received from RAN, our study focuses on the functionality and the corresponding types of inputs and outputs (massive data collected from RAN, core network, and terminals), and on potential impacts on existing nodes and interfaces; the detailed AI/ML algorithms are out of RAN3 scope. C-RAN is the separation of the RAN baseband software and the RAN baseband hardware. This baseband software can run on any capable commercial off-the-shelf (COTS) hardware, with or without integrated accelerators, utilizing cloud-native tools and processes to manage the software and hardware.

Within the RAN architecture defined in RAN3, this study prioritizes NG-RAN, including Evolved-Universal Terrestrial Radio Access New Radio Dual Connectivity (EN-DC). In terms of use cases, the group has agreed to start with energy saving, load balancing, and mobility optimization. Although the importance of avoiding duplication of SON has been recognized, additional use cases may be discussed as the study progresses, according to companies' contributions. The aim is to define a framework for AI/ML within the current NG-RAN architecture, and the AI/ML workflow being discussed should not prevent "thinking beyond," if a use case requires so.

In-Coverage Mode, 5G networks provide higher data transfer speeds by pairing a 5G RAN with the Long-Term Evolution (LTE) Evolved Packet Core (EPC). Note that the 5G RAN remains reliant on the 4G core network to manage control and signaling information and the 4G RAN continues to

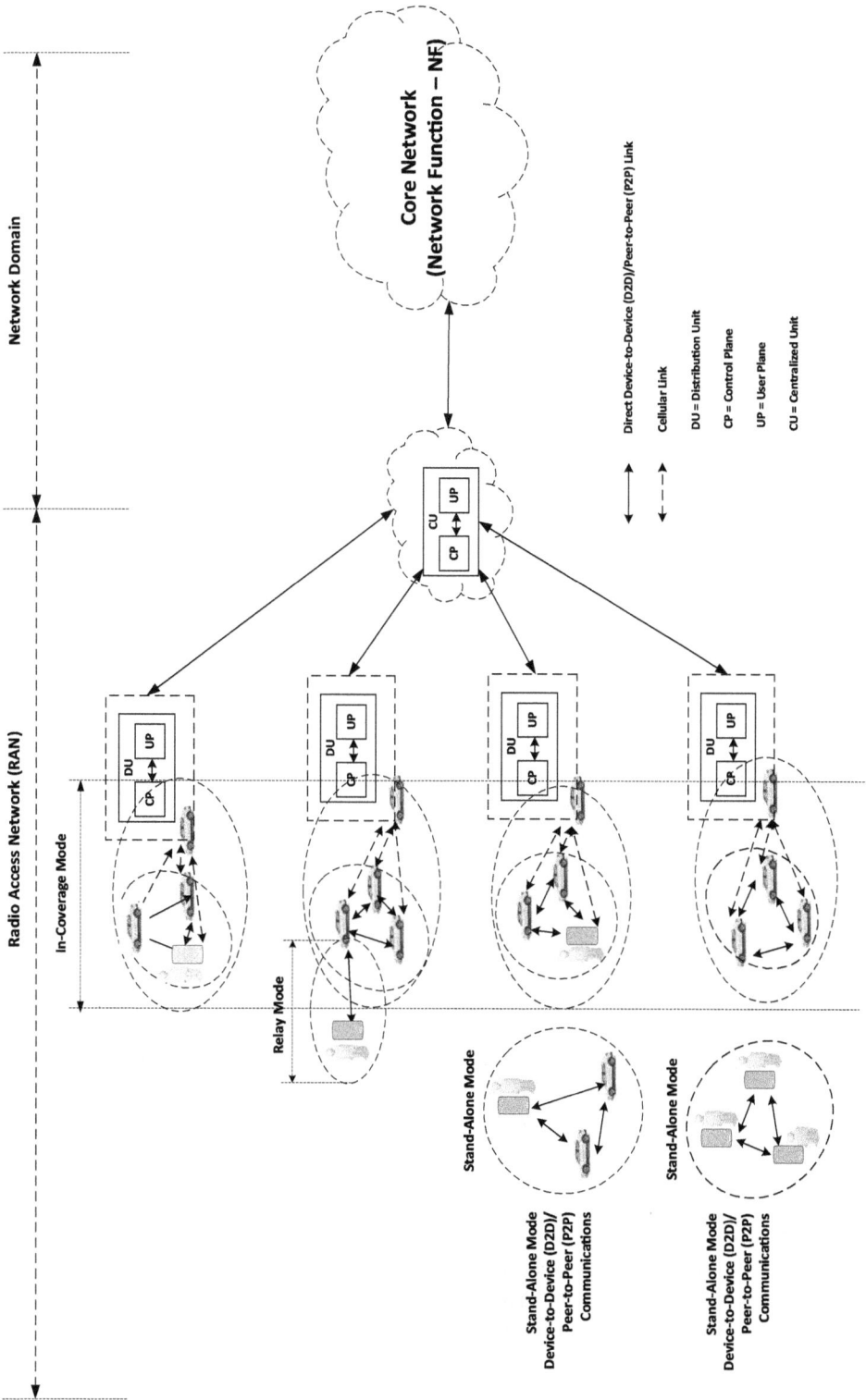

FIGURE 18.2 5G RAN and Core Network.

operate, this is called a Non-Standalone Architecture. By leveraging the existing infrastructure of a 4G network, network carriers can provide faster and more reliable Enhanced Mobile Broadband (eMBB) without completely reworking their core network technology and pushing customers to new devices. Non-Standalone 5G provides a transitionary platform for carriers and customers alike.

A massive deployment of 5G relay nodes is also considered for enhancing network capacity, coverage, and energy efficiency, and thereby there can be many variations of 5G relay networks deployment. Power supply problems need to be tackled wisely for relay networks. For example, a wirelessly powered 28-GHz phased-array relay transceiver is proposed for 5G network, which can work without any external power supply.

On the other hand, standalone 5G brings faster, more reliable, and vastly more capable telecommunications than ever before. Logistical, financial, and operational challenges of the standalone mode network operators will cost time and treasure upgrading infrastructure with 5G technologies. The standalone mode provides direct device to-device (D2D) communications which is known as dynamic peer-to-peer (P2P) communications as opposed to client-server (C/S) mode. This is a powerful concept of 5G network architecture because it does not depend on an LTE EPC to operate.

18.3   5G RADIO INTERFACE

The 5G New Radio (NR) has been developed to provide a significant enhancement in areas like flexibility, scalability, and efficiency, both in terms of power usage and spectrum. The 5G New Radio can provide communications for very high bandwidth transmissions like streaming video as well as low latency communications for remote control vehicle communications as well as low data rate low bandwidth communications for machine type communications. There are several cornerstones to the new radio used for 5G:

- **New Radio Spectrum:** Mobile communications usage is rapidly increasing, and the introduction of 5G will accelerate this trend with many more applications being accommodated by the technology. While improvements in spectrum efficiency will be made, these will not be able to accommodate the huge increases in usage, so more spectrum is needed. Release 15 also outlines several groups of new spectrums specifically for NR deployments. These range in frequency from 2.5 GHz to 40 GHz. Two bands being targeted for more immediate deployment are in the regions of 3.3 GHz to 3.8 GHz and 4.4 GHz to 5.0 GHz. The 3.3 GHz to 3.8 GHz spectrum has already been released in the US, Europe, and certain Asian countries, and they could see deployment as early as 2018. Other higher-frequency bands but below 40 GHz are also being reserved for 5G, but this is only the beginning as there is talk of usage of frequencies up to 86 GHz. The advantage of the higher frequency bands is that they are much wider, and they will be able to allow much higher signal bandwidths and hence support much higher data throughput rates. The disadvantage in some respects is that they will have a much shorter range, but this is also an advantage because it will also allow much greater frequency reuse.
- **Optimized OFDM:** An early decision was taken to use a form of OFDM as the waveform for phase one of the 5G New Radio. It has been very successfully used with 4G, the more recent Wi-Fi standards, and many other systems and came out as the optimum type of waveform for the variety of different applications for 5G. With the additional processing power available for 5G, various forms of optimization can be applied.
- The specific version of OFDM used in 5G NR downlink is cyclic prefix OFDM, CP-OFDM, and it is the same waveform LTE has adopted for the downlink signal.
- **Beamforming:** Beamforming is a technology that has become a reality in recent years, and it offers to provide some significant advantages to 5G. Beamforming enables the beam from the base station to be directed toward the mobile. In this way, the optimum signal can be transmitted to the mobile and received from it, while also cutting interference to

other mobiles. The move to higher frequencies allows for much smaller antennas and the possibility of programmable high directivity levels. On frequencies above 24 GHz where antennas are smaller, there is the possibility of having high performance beam steering antennas that are able to accurately direct the power to the mobile in question and also provide receiver gain in this direction.

- **MIMO:** MIMO, multiple input multiple output, has been employed in many wireless systems from Wi-Fi to the current 4G cellular system, and it provides some significant improvements. Within 5G, MIMO will be one of the mainstay technologies. 5G will take full advantage of Multi-User MIMO (MU-MIMO) where it will provide multiple access capabilities to MIMO by utilizing the distributed and uncorrelated spatial location of the various users.
- **5G Base Station (gNB):** In implementing this, the gNB (5G base station) sends a CSI-RS (Channel State Information Reference Signal) to the different user equipment, and then dependent upon the responses, the gNB computes the spatial information for each user. It uses this information to compute the required information for the pre-coding matrix (W-Matrix), where the data symbols are constructed into the signals for each of the elements of the gNB antenna array.
- **Multiple Data Streams:** The multiple data streams have their own weightings which include phase offsets to each stream to enable the waveforms to interfere constructively at the receiver. This maximizes the signal strength to the user while also minimizing the signal and hence interference to other users. In this way, the gNB is able to talk to multiple devices concurrently and independently by using spatial information. This means that 5G MU-MIMO enables the UEs to operate without need for knowledge of the channel or additional processing to obtain the data streams. MU-MIMO on the downlink significantly improves the capacity of the gNB antennas. It scales with the minimum of the number of gNB antennas and the sum of the number of user devices multiplied by the number of antennas per UE device. This means that using 5G MU-MIMO the system can achieve capacity gains using gNB antenna arrays and much simpler UE devices.
- **Spectrum Sharing Techniques:** Much of the radio spectrum, although allocated, is not used in an efficient manner. One of the techniques being proposed is for spectrum sharing.
- **Unified Design across Frequencies:** With the 5G New Radio utilizing a wide variety of frequencies, possibly 3.4 to 3.6 GHz below 6GHz and then 24.25 to 27.5 GHz, 27.5 to 29.5 GHz, 37 GHz, 39 GHz, and 57 to 71 GHz range as possibilities for the mm-Wave radio. It is important to have a common interface across these frequencies.
- **Small Cells:** As network densification is required to provide the required data capability more use of small cells and small cell networks are being proposed. A small cell network is a group of low-power transmitting base stations which use millimeter waves to enhance the overall network capacity. The 5G small cell network operates by coordinating a group of small cells to share the load and reduce the difficulties of physical obstructions which become more important at millimeter waves. By utilizing these techniques and many others, the 5G New radio, 5G NR will be able to significantly improve the performance, flexibility, scalability, and efficiency of current mobile networks. In this way, 5G will be able to ensure the optimum use of the available spectrum, whether it is licensed, shared, or unlicensed, and achieve this across a wide variety of spectrum bands.
- **Downlink and Uplink:** For layer 1 (physical layer), downlink (DL), that is, network to UE, NR uses OFDM with Cyclic Prefix (CP) (like LTE). For the uplink (UL), i.e., UE to network, OFDM can also be used, as well as DFT-with-OFDM (OFDM with Discrete Fourier Transform precoding). DFT-with-OFDM improves UL coverage, but it has lower peak-to-average power ratio (PAPR) and is limited to single-layer transmission only.
- **5G Layer 1 (Physical layer) Characteristics:** Some key characteristics of 5G Layer 1 is that it spreads over multiple frequency ranges to enable deployment in frequencies on a

per-country or per-region basis. The carriers are from 400 MHz up to 100 GHz, but the licensed bands are from 600 MHz up to 39 GHz. These frequencies are reframed analogue TV (UHF) bands and some satellites systems, without interference since used in different locations.

18.4   5G CORE NETWORK

Moreover, we have functionality split in 5G RAN known as vertical split control plane (CP) and user plane (UP) while the horizontal split consists of centralized unit (CU) and decentralized unit (DU). CP/UP enables the implementation of software-defined network (SDN), capacity optimization, consistent control in multi-vendor networks including interference management, and cost savings. CU/DU provides centralized resource and performance management, shifting functions to different locations based on requirements with the help of multi-access edge computing (MEC) and adaptation of RAN processing to different deployments of infrastructures. The 5G core network function (NF) is also modularized splitting into functions as depicted in Figure 18.3.

5G core network functions are defined as follows:

- **Access and Mobility Management Function (AMF):** Supports termination of non-access-stratum (NAS) signaling, NAS ciphering & integrity protection, registration management, connection management, mobility management, access authentication and authorization, security context management.
- **Session Management Function (SMF):** Supports session management (session establishment, modification, release), UE IP address allocation & management, dynamic host configuration protocol (DHCP) functions, termination of NAS signaling related to session management, DL data notification, traffic steering configuration for UPF for proper traffic routing.
- **User Plane Function (UPF):** Supports packet routing & forwarding, packet inspection, QoS handling, acts as external PDU session point of interconnect to Data Network (DN), and is an anchor point for intra- & inter-radio access technology (RAT) mobility. Note that UPF deals with anchor point for intra- and inter-RAN mobility, packet routing and forwarding, QoS handling, and traffic accounting and reporting.
- **Policy Control Function (PCF):** Supports unified policy framework, providing policy rules to CP functions, access subscription information for policy decisions in UDR.
- **Authentication Server Function (AUSF):** Acts as an authentication server.
- **Unified Data Management (UDM):** Supports generation of Authentication and Key Agreement (AKA) credentials, user identification handling, access authorization, subscription management.
- **Application Function (AF):** Supports application influence on traffic routing, accessing NEF, interaction with policy framework for policy control.
- **Network Exposure Function (NEF):** Supports exposure of capabilities and events, secure provision of information from external application to 3GPP network, translation of internal/external information.
- **NF Repository Function (NRF):** Supports service discovery function, maintains NF profile and available NF instances.
- **Network Slice Selection Function (NSSF):** Supports selecting of the Network Slice instances to serve the UE, determining the allowed NSSAI, determining the AMF set to be used to serve the UE.
- **Service Communication Proxy (SCP):** Supports traffic distribution schemes such as round robin based on capacity and its availability.
- **Network Slice Specific Authentication and Authorization Function (NSSAAF):** Specifies whether for the network slice, devices need to be also authenticated and authorized by an

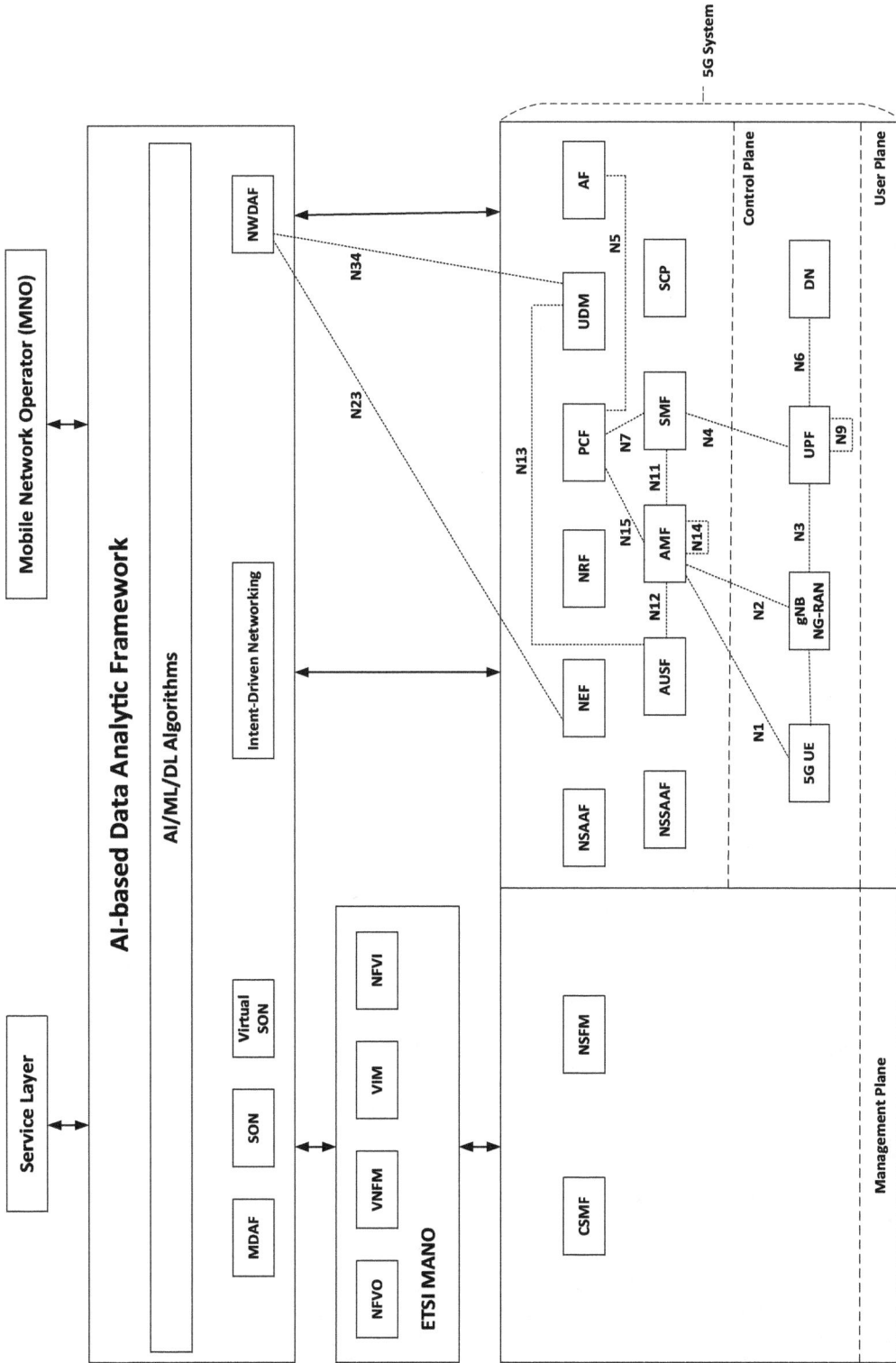

FIGURE 18.3 5G Core Network Functions (Non-Roaming Part).

Authentication, Authorization, and Accounting (AAA) server using additional credentials different than the ones used for the primary authentication.
- **Data Network (DN) Interface of 5G:** Provides connectivity to external Service Provider services, Internet access, or 3rd party services.

To deploy them, it is necessary to fully understand the customer requirements. The Communication Service Management Function (CSMF) is an entity that has this task, acting as a gateway to translate these requirements toward the network slicing ecosystem. Moreover, Network Slice Management Function (NSMF) is responsible for the management and orchestration of network slice instant (NSI) and derive network slice subnet requirements. The Network slicing management functions such as the Network Slice Management Function (NSMF), Network Slice Subnet Management Function (NSSMF), and Communication Service Management Function (CSMF) are responsible for the end-to-end creation, management, and orchestration of network slice instance, network slice subnet.

Management and Orchestration (MANO) function is a key element of the European Telecommunications Standards Institute (ETSI) network functions virtualization (NFV) architecture. MANO is an architectural framework that coordinates network resources for cloud-based applications and the lifecycle management of virtual network functions (VNFs) and network services. Finally, to orchestrate all of these, NFV MANO is built upon three functional blocks:

- **NFV Orchestrator (NFVO)**—Onboards new Network Service (NS), VNFFGs, and VNF packages. Authorizes and validates NFVI resource requests. Along with manages the NS lifecycle and performs validation/authorization of NFVI requests.
- **VNF Manager (VNFM)**—Responsible for the life cycle of VNFs, such as the creation and termination, along with the FCAPS (Fault, Configuration, Accounting, Performance and Security Management).
- **Virtualized Infrastructure Manager (VIM)**—Controls and manages the Network Function Virtualization Infrastructure (NFVI) such as the virtual/physical compute, storage, and networking resources. Along with collecting events and performance metrics.

Network Data Analytics Function (NWDAF) enables the network operators to either implement their own machine learning (ML)-based data analytics methodologies or integrate third-party solutions to their networks. NWDAF is used to bring the benefits of data-driven techniques to fruition. Analyses of the network-generated data explore core intra-network interactions through unsupervised learning, clustering, and evaluating these results as insights for future opportunities and works.

Intent-Driven Networking is an expression of the desired state of a system used to describe an intended network or service. Intents do not define specific network or service configuration, nor do they prescribe management tasks to be performed by a system. Intents allow customers to request networks and services without detailed knowledge of how they will be provided. Inherently, this assumes that the system can learn the behavior of networks and services and use AI and automation mechanisms to fulfil the requests expressed via intents. This not only relieves the consumer of the burden of knowing implementation details but also provides flexibility allowing the producer to explore alternative options to find optimal solutions.

Management Data Analytics Function (MDAF) performs the required data analysis (using the exposure capability) to help management system set rational network topology parameters for network configuration and guarantee service quality. After the rational configuration of network according to the analysis results provided by MDAF, the control plane and user plane can conduct further parameter adjustment to improve user experience. AI/ML will be used across 5G system, including management and orchestration, core network, and RAN. To enable and facilitate AI/ML operation in 5G system, AI/ML models will be created, trained, tested, deployed, and managed during its entire lifecycle. Besides, AI/ML management capabilities will coordinate with AI/ML capabilities in core network and support AI/ML capabilities in RAN.

Self-Organizing Network (SON) technology minimizes the lifecycle cost of running a mobile network by eliminating manual configuration of network elements at the time of deployment right through to dynamic optimization and troubleshooting during operation. Besides improving network performance and customer experience, SON can significantly reduce the cost of mobile operator services, improving the operation expenditure (OpEx)-to-revenue ratio and deferring avoidable capital expenditure (CapEx). SON is focusing on integrating new capabilities such as self-protection against digital security threats and self-learning through AI techniques, as well as extending the scope of SON beyond the RAN to include both mobile core and transport network segments—which will be critical to address 5G requirements such as end-to-end network slicing.

Perhaps, the most obvious candidate for AI/ML in RAN is Self-Organizing Networks (SON) functionality, currently part of LTE and NR specifications (it was initially introduced in Rel-8 for LTE). With SON, the network self-adjusts and fine-tunes a range of parameters according to the different radio and traffic conditions, alleviating the burden of manual optimization for the operator. While the algorithms behind SON functions are not standardized in 3GPP, SON implementations are typically rule-based. One of the main differences between SON and an AI-based approach is the switch from a reactive paradigm to a proactive one.

Virtual SON platforms will be developed using NFV and software-defined network (SDN) architectures. In fact, has been one of the key drivers for SON deployment in 5G networks as well as the binding of SON with C-RAN. This can occur either in a centralized or in a distributed/hybrid architecture approach, enabling the provision of SONaaS (SON as a service).

AI-based Data Analytic holds the key to revealing the true potential of the telecom industry. Traffic generated by the Internet-of-Things (IoT) might well exceed 100 trillion gigabytes in the next two years. More data means more opportunities to predict future events. With 5G's hyper-connectivity and lightning-fast speeds, AI-driven network optimization, and data analytics insights, the industry is on the cusp of a transformative revolution. The telecom sector is fluctuating as a result of the incorporation of 5G, AI, and data analytics. It delivers amplified security, hyperconnectivity, quicker speeds, intelligent network management, tailored services, and optimum resource allocation. This combination creates a data-driven, networked future that transforms user experiences and boosts economic growth. NWDAF represents the mobile industry's first attempt to standardize the function of analytics in the mobile core network. Essentially, with NWDAF, the 3GPP is ensuring that analytics are not an afterthought, and that analytics-driven automation is built into the system architecture. Some examples for NWDAF-based data analytics for application of AI/ML/DL operation are as follows:

- 5G Core Information Exposure to authorized 3rd party for Application Layer AI/ML Operation.
- Enhancing External Parameter Provisioning.
- 5G Core Enhancements to enable Application AI/ML Traffic Transport.
- QoS and Policy enhancements Existing QoS monitoring mechanisms for URLLC services are re-used.
- SMF sends the info to UPF to bind the AI/ML traffic to distinct QoS flow.
- 5G System Assistance to Federated Learning Operation.
  - New member selection functionality is added that can provide a list of recommended UEs based on the parameters contained in the request from the AF.
  - UE triggers split inference.
  - UE and an AI/ML server negotiate over application layer for the split inference operations.
- The AF subscribes to one or more NWDAF analytics to generate the AI/ML Assistance information.

- AI/ML Assistance information from 5G Core is needed to assist the AI/ML server to make corresponding decisions for the split inference.
- Based on the assistance information and the current system environmental factors such as communications data rate and resource at UE, the AI/ML server AF makes decisions on split inference.

Federated learning enables machine learning (ML) models locally trained using private data to be aggregated into a global model. Split learning allows different portions of an ML model to be collaboratively trained on different workers in a learning framework.

18.5   5G PROTOCOL STACK

The following terminologies are used per 3GPP's proposed 5G protocol stack.

- NAS = Non-Access-Stratum
- NAS-SM = NAS Session Management
- MM = Mobility Management
- 5G-AN = 5G Access Network
- NG-AP = Next Generation Application Protocol
- NR = New Radio
- gNB= gNodeB: Nodes specific to the 5G RAN base station are called gNBs.

gNodeB (gNB) is the 5G radio base station that connects 5G New Radio (NR) devices (e.g., 5G phones) to the 5G core network using the NR radio interface. 5G networks use a new technology called New Radio (NR) as described earlier, which requires new radio base stations termed as gNodeB or gNB. In the context of 5G networks, two types of radio base station nodes are described.

18.5.1   5G RADIO NETWORK NODES VS 5G DEPLOYMENT TYPES

5G network deployment can take place in two ways: standalone (SA) and non-standalone (NSA). The difference between the two deployments is the type of core network used. A non-standalone 5G deployment makes use of the 4G core network, whereas a standalone deployment uses a dedicated 5G core network. The non-standalone mode (NSA) is the more common deployment option, especially for the early adoption of 5G, where mobile operators add 5G NR to their existing 4G LTE infrastructure. NSA makes use of the LTE core network EPC (with some enhancements), whereas SA uses a new 5G core network called 5G Core Network.

18.5.2   5G USER AND CONTROL PLANES FOR eNB, gNB, AND NG-eNB

With a 5G core network, gNB is responsible for both user and control planes for 5G NR devices, with a 4G core network in non-standalone 5G, gNB is responsible for the user plane, and eNB is responsible for the control plane. The dual connectivity concept is a significant one for the coexistence of LTE and 5G. In dual connectivity, a mobile phone is connected to both 4G and 5G networks simultaneously; however, the type of connectivity is different. In a non-standalone deployment (NSA), the size of the data connection, quality of service (QoS), and other user-level functions for a 5G NR device (e.g., a 5G phone) are handled by the 5G NR radio network base station (gNodeB or gNB). This user-level connection is called the user plane. However, the control-level functions for the 5G device are part of the control plane and are handled by the 4G LTE radio network termed as evolved node base station (eNodeB or eNB).

When a 5G core network is used, the connectivity is different. When a 5G device needs to connect to the network, the gNodeB (gNB) can provide both control and user planes using the NR radio interface. On the other hand, when a 4G device needs to connect to the network, next generation evolved node base station (ng-eNodeB or ng-eNB) can provide both control and user planes using the LTE radio interface.

18.5.3   5G Control Plane: UE-to-AMF and UE-to-SMF Protocol Stack

A protocol stack is defined, e.g., in TS 23.501 for communications between several of these NFs, and secondary ones, not presented in Figure 18.4. Here, we highlight some of the main ones:

- **NAS-SM**: It supports the handling of Session Management between the UE and the SMF. It supports user plane PDU Session establishment, modification, and release. It is transferred via the AMF and transparent to the AMF. It is defined in "Non-Access-Stratum (NAS) protocol for 5G System (5GS); Stage 3" (TS 24.501).
- **NAS-MM**: It supports registration management functionality, connection management functionality, and user plane connection activation and deactivation. It is also responsible for ciphering and integrity protection of NAS signaling. 5G NAS protocol is defined in TS 24.501.
- **5G-AN Protocol Layer**: This set of protocols/layers depends on the 5G-AN. In the case of NG-RAN, the radio protocol between the UE and the NG-RAN node (eNodeB or gNodeB) is specified in the E-UTRA & E-UTRAN; "Overall description; Stage 2" (TS 36.300) and the NR "Overall description; Stage-2" in TS 38.300. In the case of non-3GPP access, see clause 8.2.4.
- **NG Application Protocol** (NG-AP): Application Layer Protocol between the 5G-AN node and the AMF. NG-AP is defined in TS 38.413.
- **Stream Control Transmission Protocol (SCTP)**: This protocol guarantees delivery of signaling messages between AMF and 5G-AN node (N2). SCTP is defined in IETF RFC 4960 [1].

Note that there is also a direct communication between 5G-AN and SMF, called N2 SM information: this is the subset of NG-AP information (not shown in the figure) that the AMF transparently relays between the 5G-AN and the SMF and is included in the NG-AP messages and the N11 related messages.

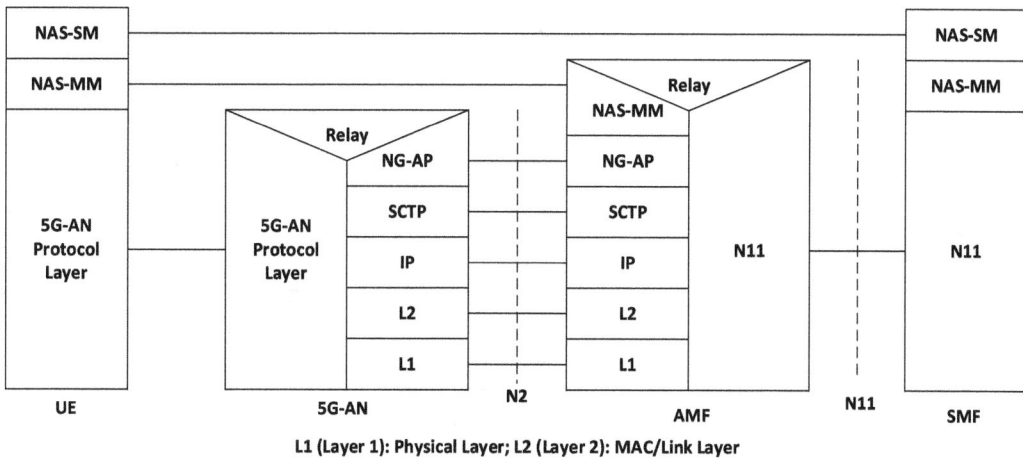

L1 (Layer 1): Physical Layer; L2 (Layer 2): MAC/Link Layer

FIGURE 18.4   5G Protocol Stack: Control Plane Protocol Stack between UE, 5G-AN, AMF, and SMF.

18.5.4   5G User Plane: UE-to-AMF and UE-to-SMF Protocol Stack

Figure 18.5 is extracted from TS 23.501, section 8.3. It illustrates the protocol stack for the User plane transport related with a PDU Session.

- **PDU Layer**: This layer corresponds to the PDU carried between the UE and the DN over the PDU Session. When the PDU Session Type is IPv4 or IPv6 or IPv4v6, it corresponds to IPv4 packets or IPv6 packets or both; When the PDU Session Type is Ethernet, it corresponds to Ethernet frames; others.
- **GPRS Tunnelling Protocol for the User Plane (GTP U)**: This protocol supports tunneling user data over N3 (i.e., between the 5G-AN node and the UPF) and N9 (i.e., between different UPFs of the 5GC) in the backbone network, for details, see TS 29.281. GTP shall encapsulate all end user PDUs. It provides encapsulation on a per PDU Session level. This layer carries also the marking associated with a QoS Flow defined in clause 5.7. This protocol is also used on N4 interface as defined in TS 29.244.
- **5G-AN Protocol Stack**: This set of protocols/layers depends on the AN. When the 5G-AN is a 3GPP NG-RAN, these protocols/layers are defined in TS 38.401. The radio protocol between the UE and the 5G-AN node (eNodeB or gNodeB) is specified in TS 36.300 and TS 38.300. L2 is also called the "Data Link Layer" and the L1 is the "Physical Layer".
- **UDP/IP**: These are the backbone network protocols.

18.5.5   5G Non-Standalone versus 5G Standalone Architecture

Two deployment options are defined for 5G: the "Non-Stand Alone" (NSA) architecture, where the 5G Radio Access Network (AN) and its New Radio (NR) interface is used in conjunction with the existing LTE and EPC infrastructure Core Network (respectively 4G Radio and 4G Core), thus making the NR technology available without network replacement. In this configuration, only the 4G services are supported, but they enjoy the capacities offered by the 5G New Radio (lower latency, and others). The NSA is also known as "E-UTRA-NR Dual Connectivity (EN-DC)" or "Architecture Option 3." The NSA architecture is illustrated in Figure 18.6.

The NSA architecture can be seen as a temporary step toward a "full 5G" deployment, where the 5G Access Network is connected to the 4G Core Network. In the NSA architecture, the (5G) NR base station (logical node "en-gNB") connects to the (4G) LTE base station (logical node "eNB")

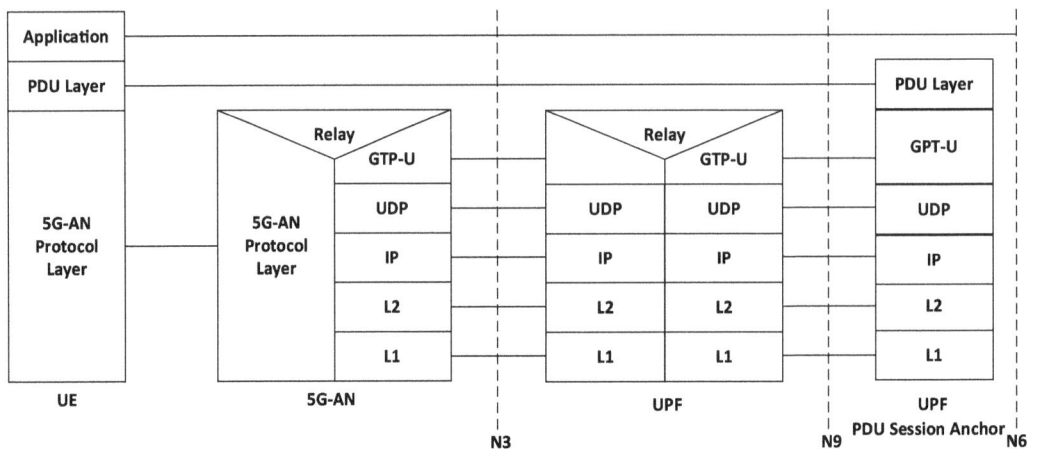

FIGURE 18.5   User Plane Protocol Stack between UE, 5G-AN and UPF.

via the X2 interface. The X2 interface was introduced prior to Release 15 to connect two eNBs. In Release 15, it also supports connecting an eNB and en-gNB to provide NSA. The NSA offers dual connectivity, via both the 4G AN (E-UTRA) and the 5G AN (NR). It is thus also called "EN-DC," for "E-UTRAN and NR Dual Connectivity." In EN-DC, the 4G's eNB is the Master Node (MN) while the 5G's en-gNB is the Secondary Node (SN).

The "Stand-Alone" (SA) architecture is where the NR is connected to the 5G Core Network. Only in this configuration, the full set of 5G Phase 1 services are supported. The SA architecture is illustrated in Figure 18.7.

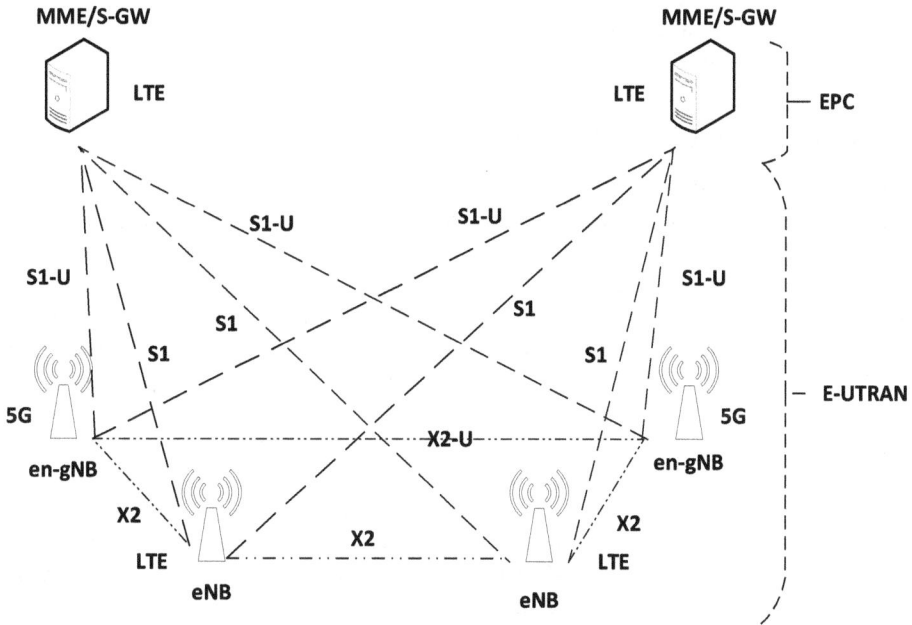

FIGURE 18.6 The NSA Architecture.

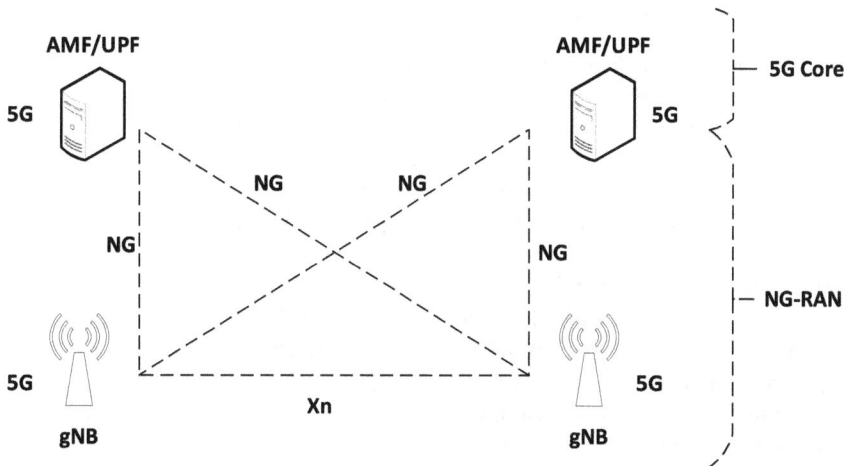

FIGURE 18.7 The SA Architecture.

The SA architecture can be seen as the "full 5G deployment," not needing any part of a 4G network to operate. The NR base stations (logical node "gNB") connect with each other via the Xn interface, and the Access Network (called the "NG-RAN for SA architecture") connects to the 5G Core network using the NG interface.

18.6   5G SECURITY

The 5G system is an evolution of the 4G mobile communication systems. Accordingly, the 5G security architecture is designed to integrate 4G equivalent security for both non-Standalone and the Standalone new radio (NR) 5G systems. In addition, the reassessment of other security threats such as attacks on radio interfaces, signaling plane, user plane, masquerading, privacy, replay, bidding down, man-in-the-middle, and inter-operator security issues have also been taken into account for 5G and will lead to further security enhancements.

18.6.1   5G Non-Standalone NR Security

The first step standardized by 3GPP toward complete 5G coverage was Non-Standalone NR, also known as E-UTRA-NR Dual Connectivity (EN-DC). The key feature of non-standalone is the ability to utilize existing LTE and EPC infrastructure, thus making new 5G-based radio technology available without network replacement. EN-DC uses LTE as the master radio access technology while the new radio access technology (i.e., NR) serves as secondary radio access technology with User Equipment (UE) connected to both radios. Except for capability negotiation, security procedures for EN-DC basically follow the specifications for dual connectivity security for 4G.

A master evolved node base station eNB (MeNB) checks whether the UE has 5G NR capabilities to access the Secondary gNB (SgNB), i.e., 5G base-station, and the access rights to SgNB. The capability and access rights check ensures that the standard is forward compatible since UEs with different capabilities, including security capabilities, can join the network. The MeNB derives and sends the key to be used by the SgNB for secure communication over NR; the UE also derives the same key. Unlike dual connectivity in 4G networks, Radio Resource Control (RRC) messages can be exchanged between the UE and SgNB, thus keys used for integrity and confidentiality protection of RRC messages as well as User Plane (UP) data are derived. Although integrity protection for UP data is supported in 5G network, it will not be used in EN-DC case. The use of confidentiality protection is optional for both UP and RRC.

18.6.2   Evolution of the 5G Trust Model

Moving on from the non-Standalone deployment, in a Standalone 5G system, the trust model has evolved. Trust within the network is considered as decreasing the further one moves from the core. This has an impact on decisions taken in 5G security design. The trust model in the UE is reasonably simple: there are two trust domains, the tamper proof universal integrated circuit card (UICC) on which the Universal Subscriber Identity Module (USIM) resides as trust anchor and the Mobile Equipment (ME). The ME and the USIM together form the UE. The network side trust model for roaming and non-roaming cases are shown in Figures 18.8 and 18.9 respectively, which shows the trust in multiple layers, like in an onion.

The Radio Access Network (RAN) is separated into distributed units (DU) and central units (CU)—DU and CU together form gNB 5G base-station. The DU does not have any access to customer communications as it may be deployed in unsupervised sites. The CU and Non-3GPP Inter Working Function (N3IWF—not shown in the figures), which terminates the Access Stratum (AS) security, will be deployed in sites with more restricted access.

In the core network the Access Management Function (AMF) serves as termination point for Non-Access Stratum (NAS) security. Currently, i.e., in the 3GPP 5G Phase 1 specification, the

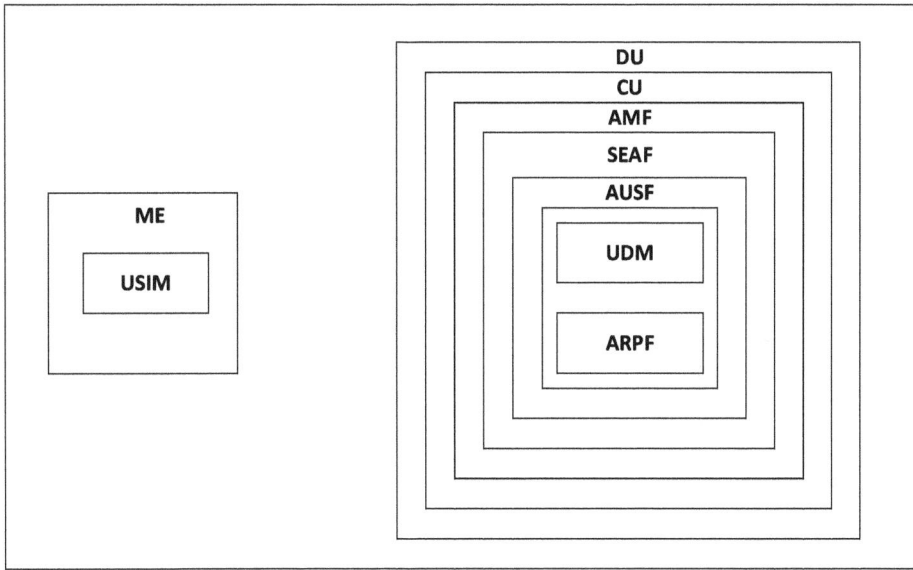

FIGURE 18.8 Trust Model of Non-Roaming Scenario.

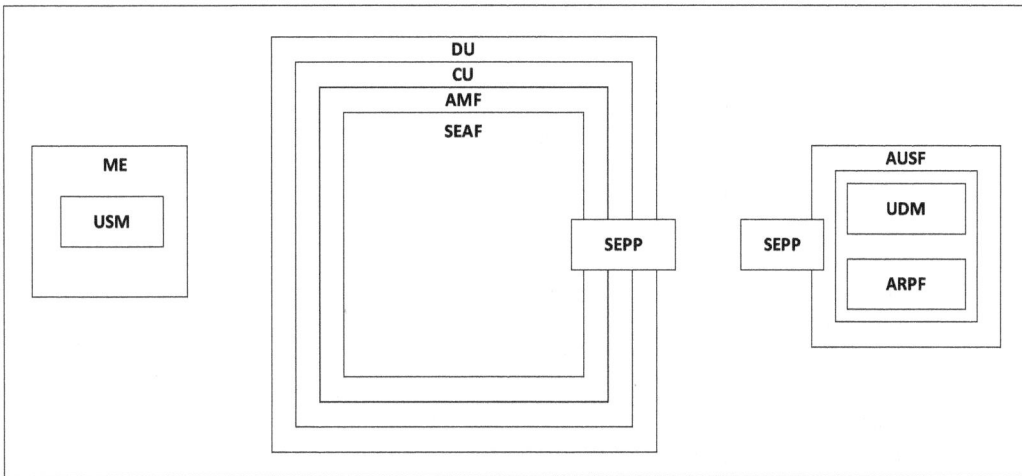

FIGURE 18.9 Trust Model of Roaming Scenario.

AMF is collocated with the SEcurity Anchor Function (SEAF) that holds the root key (known as anchor key) for the visited network. The security architecture is defined in a future-proof fashion, as it allows separation of the security anchor from the mobility function that could be possible in the future.

Authentication Function (AUSF) keeps a key for reuse, derived after authentication, in case of simultaneous registration of a UE in different access network technologies, i.e., 3GPP access networks and non-3GPP access networks such as IEEE 802.11 Wireless Local Area Network (WLAN). Authentication credential Repository and Processing Function (ARPF) keeps the authentication credentials. This is mirrored by the USIM on the side of the client, i.e., the UE side. The subscriber information is stored in the Unified Data Repository (UDR). The Unified Data Management (UDM) uses the subscription data stored in UDR and implements the application logic to perform various functionalities such as authentication credential generation, user identification, service, and session continuity, and so on. Over the air interface, both active and passive attacks are considered on both control plane and user plane. Privacy has become increasingly important leading to permanent identifiers being kept secret over the air interface.

In the roaming architecture, the home and the visited network are connected through SEcurity Protection Proxy (SEPP) for the control plane of the internetwork interconnect. This enhancement is done in 5G because of the number of attacks coming to light recently such as key theft and re-routing attacks in SS7 and network node impersonation and source address spoofing in signaling messages in DIAMETER [2] that exploited the trusted nature of the internetwork interconnect.

18.6.3   5G Phase 1 Security (Release 15)

5G Phase 1 brings several enhancements to 4G LTE security, some of the key points are presented in this section. Details of 5G Phase 1 security are briefly described here.

- **Primary Authentication**: Network and device mutual authentication in 5G is based on primary authentication. This is like 4G, but there are a few differences. The authentication mechanism has in-built home control allowing the home operator to know whether the device is authenticated in each network and to take the final call of authentication. In 5G Phase 1, there are two mandatory authentication options: 5G Authentication and Key Agreement (5G-AKA) and Extensible Authentication Protocol (EAP)-AKA', i.e., EAP-AKA'. Optionally, other EAP-based authentication mechanisms are also allowed in 5G—for specific cases such as private networks. Also, primary authentication is radio access technology independent, thus it can run over non-3GPP technology such as IEEE 802.11 WLANs.
- **Secondary Authentication**: Secondary authentication in 5G is meant for authentication with data networks outside the mobile operator domain. For this purpose, different EAP-based authentication methods and associated credentials can be used. A similar service was possible in 4G as well, but now it is integrated in the 5G architecture.
- **Inter-Operator Security**: Several security issues exist in the inter-operator interface arising from SS7 or Diameter [5,6] in the earlier generations of mobile communication systems. To counter these issues, 5G Phase 1 provides inter-operator security from the very beginning.
- **Privacy**: Subscriber identity-related issues have been known since 4G and earlier generations of mobile systems. In 5G, a privacy solution is developed that protects the user's subscription permanent identifier against active attacks. A home network public key is used to provide subscriber identity privacy.
- **Service-based Architecture (SBA)**: The 5G core network is based on a service-based architecture, which did not exist in 4G and earlier generations. Thus, 5G also provides adequate security for SBA.
- **Central Unit (CU)—Distributed Unit (DU)**: In 5G, the base-station is logically split in CU and DU with an interface between them. Security is provided for the CU–DU interface. This split was also possible in 4G, but in 5G it is part of the architecture that can support several deployment options (e.g., co-located CU–DU deployment is also possible). The DUs, which are deployed at the very edge of the network, don't have access to any user data when confidentiality protection is enabled. Even with the CU–DU split, the air interface security point in 5G remains the same as in 4G, namely in the radio access network.
- **Key Hierarchy**: The 5G hierarchy reflects the changes in the overall architecture and the trust model using the security principle of key separation. One main difference in 5G compared to 4G is the possibility for integrity protection of the user plane.
- **Mobility**: Although mobility in 5G is similar to 4G, the difference in 5G is the assumption that the mobility anchor in the core network can be separated from the security anchor.

Non-standalone and 5G Phase 1 standalone architecture gave us a taste of the new generation of the mobile communication system. The main use case for 5G Phase 1 was mobile broadband. 5G Phase 2 will bring solutions for the Internet-of-Things (IoT), covering several scenarios in the form of massive Machine Type Communication (mMTC) and Ultra-Reliable and Low Latency Communications (URLLC). mMTC relates to a very large number of devices transmitting a relatively low volume of non-delay-sensitive data, and URLLC relates to services with stringent requirements for capabilities such as throughput, latency, and availability.

For mMTC, very low data-rates going down to a few bits per day, we will have to consider the extent of security (be it authentication, confidentiality, integrity, or otherwise) that can be provided. Several IoT or Machine-to-Machine (M2M) services and devices fall under this category, examples are temperature sensors giving hourly updates, sensors on farm animals giving vital status couple of times a day, etc. Such devices will also be resource constrained in terms of battery, computation, and memory. For security, the requirement will be to reduce the overhead of security-related bits, e.g., for integrity, for every communication.

At the other end of the scale, URLLC devices will call for high data-rates with potentially higher battery and computational resources; examples include cars, Industrial IoT (IIoT) devices like factory machinery and virtual or augmented reality (VR or AR) devices—used for gaming or real-time services. Providing higher data rates also means that throughput of security functions needs to be considered to avoid processing delay.

5G end-to-end backbone network is shown in Figure 18.10. The core backbone network consists of fiber/optical/photonic and time/dense wavelength technologies. AI-driven intelligent fiber/optical/photonic and dense wavelength network offers efficiency for optimal coding, channel emulator, MIMO, channel equalization, and optimal decision in view of nonlinear complex noises of lights. Optical computing and optical neural networks are the focus of development for intelligent optical communications network (IOCN). IOCN can be deployed in 6G on a large scale, although electricity-based digital signal processing will consume a lot of energy. The use of optical computing will greatly reduce consumption. Distributed channel equalization combines the existing communication system and the multilayer mechanism of neural networks, which will be an effective means of rapid deployment of IOCN.

In AI-enabled networks, all of these protocols will be controlled by AI which, in turn, invoke a given or a set of ML/DL algorithms as appropriate dynamically for solving the problems at hand.

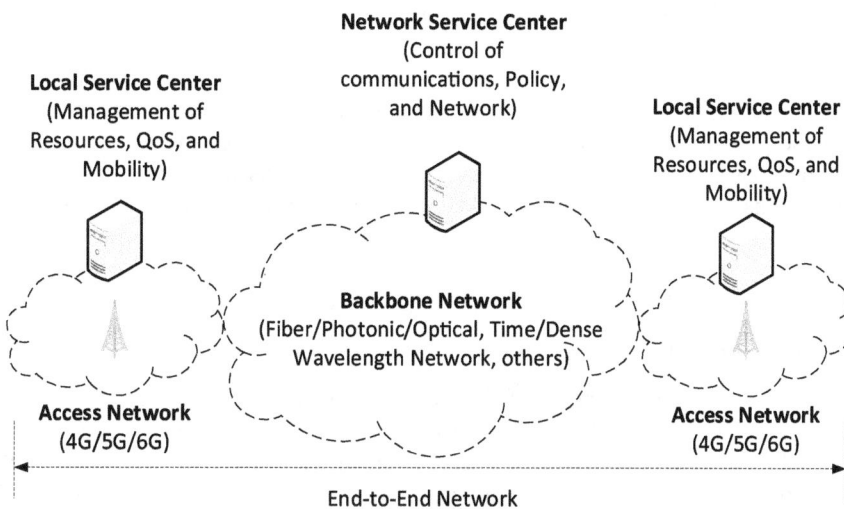

FIGURE 18.10   5G End-to-end Network.

There are a finite number of ML/DL algorithms, but their usage is countless. Many research papers have been published over the years for making each of these protocols AI-enabled. In this way, each of these protocols lose their original legacy characteristics to make them dynamic and intelligent to meet the stringent response time and hence the throughput capacity of each protocol increases almost to a level as if not so lossy network. As a result, AI-enabled networks will provide higher system capacity, higher data rate, lower latency, higher security, and improved quality of service (QoS) compared to those of the legacy non-AI-enabled network. However, standardization requires to define each of these protocols need to develop detailed specifications in open forums, IETF, ITU-T, ISO/IEC, IEEE, and others for providing interoperability in multi-vendor environments.

18.7   5G NETWORK SLICING

A key service idea of 5G is network slicing that offers extreme flexibility for service creation. For example, Slice 1 is for optimization of low latency, Slice 2 for high bandwidth optimization, Slice 3 for low power devices, and so on. The 5G standards also allow services like enhanced mobile broadband (eMBB), ultra-reliable low latency communications (URLLC), and mass machine-type communications (mMTC). An implementation scenario (Figure 18.11) could be by slicing their networks into three parts, each with its own set of network functions.

The slicing of network services enables different virtualization of services based on the needs of the customers instead of one kind of plain service like virtual private network (VPN) as we see in legacy networks. For example, customers connecting Internet-of-Thing (IoT) devices of a utility company might use mMTC service on a mass scale for monitoring its smart meters. Network service providers might use URLLC service for premium customers who demand high reliability, extremely low latency with moderate bandwidth demand while the eMBB service will be provided to their premium customers who demand high reliability, high bandwidth, and extremely low latency. The key is the applications that are used by customers because applications could be consisting of audio, video, data, graphics, and/or texts having real-time, near-real-time, or non-real-time traffic characteristics.

The further granularity of URLC, eMBB, and mMTC services can be made as depicted in Figure 18.12. In each category of services different network slices can be created customizing the services for each customer, and slices can be created according to a given slice depending on latency, bandwidth, and reliability requirements. Note that each of the URLC, eMBB, and mMTC services defines the structure, configuration, and workflows for the network slice instance throughout its lifecycle while slices (e.g., 1, 2, 3, etc.) define specific service requirements of the customer's applications.

A key paradigm shift in high bandwidth in 5G networks is the granularity services and their finer performance requirements supporting a huge customer based with massive management of mobility of the global network supporting real-time, near-real-time, and non-real-time applications. It implies that AI/ML/DL technologies are essential to be used for creation of these network services as well as for customers' applications. The 5G network is only the precursor of the 6G network where the AI/ML/DL technologies are being used for networks and applications services creation.

18.8   5G MAC PROTOCOL

The massive connection requirements of the Internet-of-Things (IoT) devices where high bandwidth is also not essential for all users has given rise to a new standardized protocol known as non-orthogonal multiple access (NOMA). There can be many varieties of schemes in NOMA protocols. The new NOMA access protocols are considered as a key enabling technique for 5G from the academia, standard bodies, and wireless industry. NOMA protocols allow for the sharing of a single resource among a number of different subscribers. These classes of protocols are considered

Ultra-Reliable Low Latency Communications (URLLC)
- High Reliability
- Medium Bandwidth
- Extremely Low Latency

Enhanced Mobile Broadband (eMBB)
- High Reliability
- High Bandwidth
- Extremely Low Latency

Mass Machine-Type Communications (mMTC)
- Low Bandwidth
- Not Sensitive or Latency and Reliability

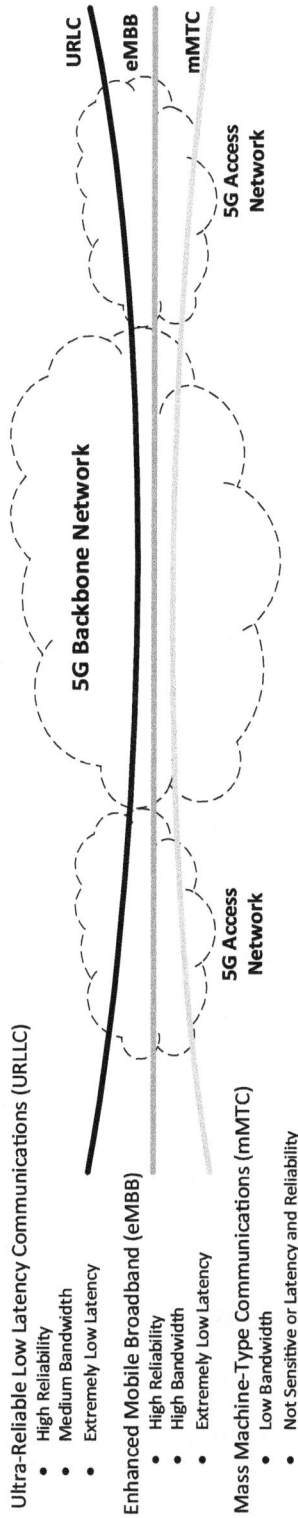

FIGURE 18.11 Slicing of Network with URLLC, eMMB, and mMTC Services.

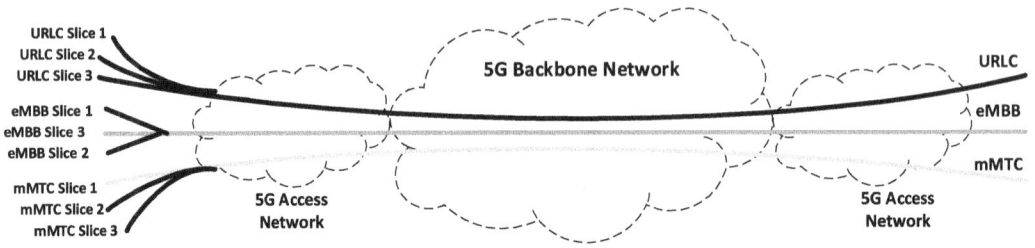

FIGURE 18.12 Granularizations of eMBB, URLLC, mMTC Services.

as a possible solution for the continuous expansion of wireless communication networks for mass connection of IoT devices. NOMA technologies can increase the effectiveness of orthogonal access systems due to their compatibility with other multiple access methods.

In NOMA, each user is given exclusive use of a separate signal or frequency range using orthogonal access. Multiple users can share the same frequency or specific time in a network that employs non-orthogonal multiple access. Furthermore, a NOMA system allows for flexible power distribution strategies to ensure a more equitable recycling of resources for users located further from the epicenter of the networks. The key driver of these new protocols is the need for increasing massive connectivity density. NOMA facilitates the efficient use of the scare resources of time and spectrum in wireless networks like 5G and could be recommended for 6G.

Note that it is desirable that once a user has data to transmit, it can start immediately without fully setting up the connection, nor relying on the dynamic grant for increasing efficiency of the channel use. In other words, a user is allowed to transmit data in radio resource controller (RRC) idle or inactive. In grant-free transmission, each user randomly selects a resource or a signature. There would be potential collision between users. NOMA is able to mitigate collision issues by applying specific transmission schemes together with advanced receivers. With NOMA, the system can withstand a greater number of active users or heavier loading and thus increase the system capacity. In mMTC, eMBB, and URLLC, NOMA schemes have been recommended.

The support of AI/ML/DL technologies needs to be used in NOMA technologies to meet performance requirements. This aspect also needs to be standardized to have interoperability in multi-vendor environments. Standard bodies and industries are yet to work in this area.

18.9 5G HIGHER LAYER PROTOCOLS

The higher-layer protocols (e.g., network, transport, and middleware) are yet to be addressed for improvement like MAC layer to have the end-to-end benefits in terms of overall performances like throughput capacity and latencies. AI-enabled higher layer protocols are essential to reap the end-to-end performance benefits.

18.10 AI-ENABLED 5G APPLICATIONS

The 5G network has been standardized for AI-enabled applications. So AI is simply an overlaid service. In this scenario, AI-based applications servers will work as the backend server to enable legacy applications or software updates of AI-enabled legacy applications. Depending on implementation, AI-enabled applications servers can be placed in the backbone and/or access network (Figure 18.13).

This AI-enabled 5G network is transparent implementations. However, AI-enabled applications will still be proprietary although frontend application protocol could be standards-based.

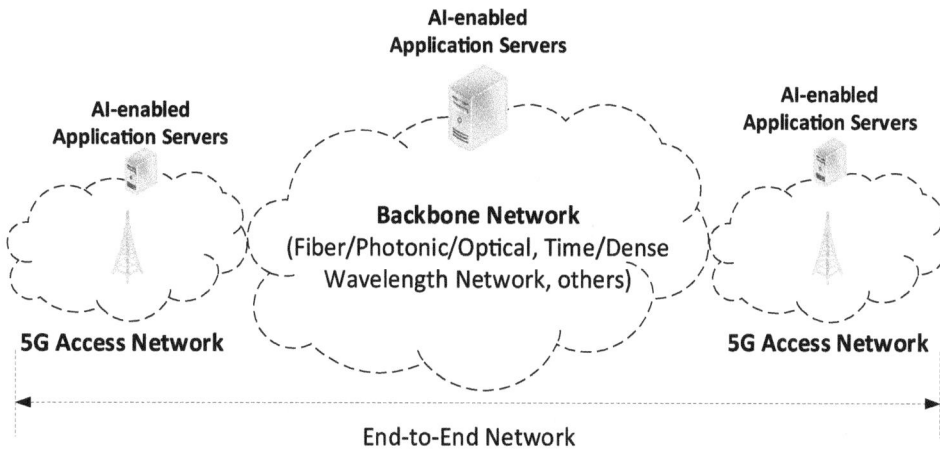

FIGURE 18.13   5G End-to-End Network.

18.11   5G END-TO-END NETWORK ARCHITECTURE WITH AI-ENABLED APPLICATIONS

5G networks will use the legacy network protocols that are not AI-enabled. As described earlier, the 5G network will be using the AI-enabled applications to enhance the existing services to be more efficient and improvising latencies. Figure 18.14 depicts the protocol architecture of the 5G network where applications are AI-enabled.

Note that 3GPP would not define exact AI/ML/DL network definition and henceforth its implementation. It would define only the interface and type of input/output of the AI/ML/DL network (model). The whole AI/ML/DL functionality would comprise several different components (e.g., Data Collection, Model Training, Model Inference, Actor). Perhaps, the most obvious candidate for AI/ML in RAN is Self-Organizing Networks (SON) functionality, currently part of LTE and NR specifications (it was initially introduced in Rel-8 for LTE). With SON, the network self-adjusts and fine-tunes a range of parameters according to the different radio and traffic conditions, alleviating the burden of manual optimization for the operator. While the algorithms behind SON functions are not standardized in 3GPP, SON implementations are typically rule-based. One of the main differences between SON and an AI-based approach is the switch from a reactive paradigm to a proactive one. Note that 3GPP defines RAN in the following groups:

- **RAN1**—Radio Layer 1: Physical layer.
- **RAN2**—Radio layer 2 and Radio layer 3 Radio Resource Control.
- **RAN3**—Universal Terrestrial Radio Access Network (UTRAN)/Evolved UTRAN (E-UTRAN)/Next Generation RAN (NG-RAN) including Cloud RAN (C-RAN) architecture and related network interfaces.
- **RAN4**—Radio Performance and Protocol Aspects.
- **RAN5**—Mobile terminal conformance testing.

The study has just begun, and at the time of writing we can only provide initial considerations. According to the mandate received from RAN, our study focuses on the functionality and the corresponding types of inputs and outputs (massive data collected from RAN, core network, and terminals), and on potential impacts on existing nodes and interfaces; the detailed AI/ML algorithms are out of RAN3 scope. C-RAN is the separation of the RAN baseband software and the RAN baseband hardware. This baseband software can run on any capable commercial off-the-shelf (COTS)

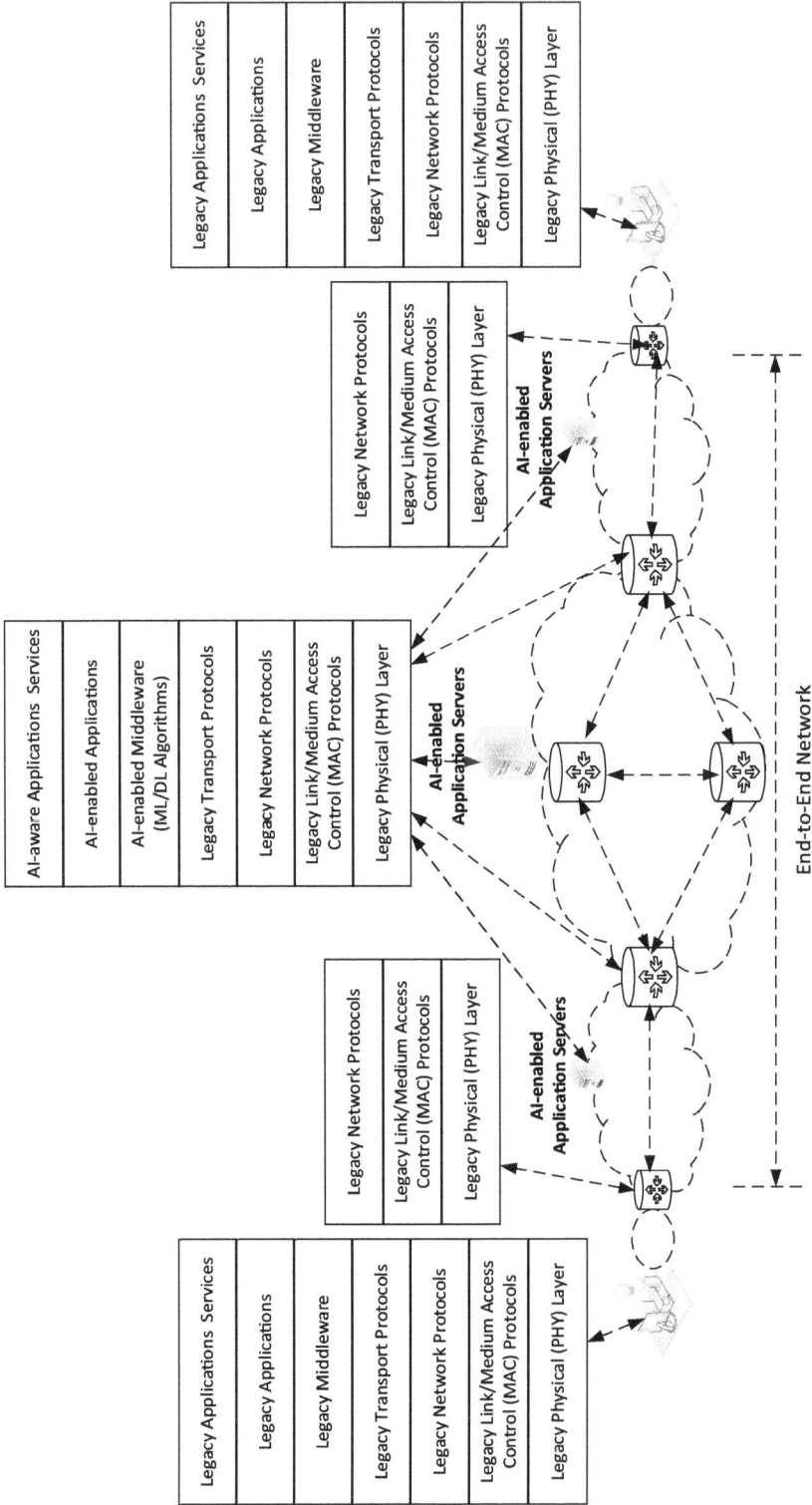

FIGURE 18.14 5G Network Architecture with AI-enabled Applications.

hardware, with or without integrated accelerators, utilizing cloud-native tools and processes to manage the software and hardware.

Within the RAN architecture defined in RAN3, this study prioritizes NG-RAN, including Evolved-Universal Terrestrial Radio Access New Radio Dual Connectivity (ENDC). In terms of use cases, the group has agreed to start with energy saving, load balancing, and mobility optimization. Although the importance of avoiding duplication of SON was recognized, additional use cases may be discussed as the study progresses, according to companies' contributions. The aim is to define a framework for AI/ML within the current NG-RAN architecture, and the AI/ML workflow being discussed should not prevent "thinking beyond," if a use case requires so.

18.12 SUMMARY

We have described an AI-enabled 5G network. 5G RAN, 5G radio Interface, 5G core network, 5G protocol stack, 5G security, 5G network slicing, 5G MAC protocol, 5G higher layer protocols, AI-enabled 5G applications, and 5G end-to-end network architecture with AI-enabled applications. All of these functionalities and capabilities are primarily explained based on 3GPP emerging specifications.

REFERENCES

[1] RFC 4850, "ICMP Extensions for Multiprotocol Label Switching," August 2007.
[2] RFC 6733, "DIAMETER Base Protocol," October 2012.

Index

For Product Safety Concerns and Information please contact our EU
representative GPSR@taylorandfrancis.com
Taylor & Francis Verlag GmbH, Kaufingerstraße 24, 80331 München, Germany